Non-Metallic Technical Textiles

This book describes various aspects of technical textiles and materials, emerging technologies, plant by-products, ultrafine fibers, functional fibers, and fabrics, covering the entire spectrum of technical textiles. It covers the fundamental aspects of emerging technology, materials, and processes. It also discusses various futuristic potential nanofibrous material spun via needleless technology and their inherent properties utilized for creating functional applications in the field of technical textiles.

Features:

- Covers the fundamentals of technical fibers and their processing technologies.
- Explores natural fibers from agro-residue for high-value technical textiles.
- Presents up-to-date summary of technical textiles and associated technology.
- Highlights research and development studies data translated into product-oriented research and practical applications.
- Identifies the coloring ability of prevailing and new sources of pigments from bioresources.

The book is aimed at researchers, professionals, and graduate students in textile and industrial engineering, materials science, and engineering, including apparel engineering.

Emerging Materials and Technologies

Series Editor: Boris I. Kharissov

The *Emerging Materials and Technologies* series is devoted to highlighting publications centered on emerging advanced materials and novel technologies. Attention is paid to those newly discovered or applied materials with potential to solve pressing societal problems and improve quality of life, corresponding to environmental protection, medicine, communications, energy, transportation, advanced manufacturing, and related areas.

The series takes into account that, under present strong demands for energy, material, and cost savings, as well as heavy contamination problems and worldwide pandemic conditions, the area of emerging materials and related scalable technologies is a highly interdisciplinary field, with the need for researchers, professionals, and academics across the spectrum of engineering and technological disciplines. The main objective of this book series is to attract more attention to these materials and technologies and invite conversation among the international R&D community.

Emerging Materials and Technologies for Bone Repair and Regeneration
Edited by Ashok Kumar, Sneha Singh, and Prerna Singh

Mechanics of Auxetic Materials and Structures
Farzad Ebrahimi

Nanomaterials for Sustainable Hydrogen Production and Storage
Edited by Jude A. Okolie, Emmanuel I. Epelle, Alivia Mukherjee, and Alaa El Din Mahmoud

Calcium-Based Materials
Processing, Characterization, and Applications
Edited by S.S. Nanda, Jitendra Pal Singh, Sanjeev Gautam, and Dong Kee Yi

Advanced Synthesis and Medical Applications of Calcium Phosphates
Edited by S.S. Nanda, Jitendra Pal Singh, Sanjeev Gautam, and Dong Kee Yi

Non-Metallic Technical Textiles
Materials and Technologies
Mukesh Kumar Sinha and Ritu Pandey

For more information about this series, please visit: www.routledge.com/Emerging-Materials-and-Technologies/book-series/CRCEMT

Non-Metallic Technical Textiles
Materials and Technologies

Mukesh Kumar Sinha and Ritu Pandey

CRC Press
Taylor & Francis Group
Boca Raton London New York

CRC Press is an imprint of the
Taylor & Francis Group, an **informa** business

Designed cover image: courtesy of Mukesh Kumar Sinha, the cover photo is the microscopic image of naturally occurring technical fabric covering the young coconut trunk. The young coconut trunk emerges wrapped up in this woven yet breathable fabric. The fabric protects the trunk from harsh external environment.

First edition published 2024
by CRC Press
2385 NW Executive Center Drive, Suite 320, Boca Raton FL 33431

and by CRC Press
4 Park Square, Milton Park, Abingdon, Oxon, OX14 4RN

CRC Press is an imprint of Taylor & Francis Group, LLC

© 2024 Mukesh Kumar Sinha and Ritu Pandey

ISBN: 9781032328614 (hbk)
ISBN: 9781032328638 (pbk)
ISBN: 9781003317074 (ebk)

DOI: 10.1201/9781003317074

Typeset in Times
by Apex CoVantage, LLC

Contents

Preface

The book is the outcome of the difficulties faced by the authors in synthesizing the subject while addressing it for professional students and researchers. As is known, the emergence of technical textiles is as old as traditional textiles. The treasures of Tutankhamun (1323 BCE) included not only the world's oldest garment, the *Tarkhan Dress*, but also linen sheets, which were used to bandage the mummies of Egyptian pharaohs. Missives used in the ancient kingdom were made from plant fibers. Metallic textiles were largely used for protection by soldiers up to the early 19th century. Modern-day protective technical textiles used by combatants are non-metallic and lightweight with features of impact and crack resistance. Over the years, use of technical textiles increased manifold with greater functionalities serving the needs of agro-, medical, packaging, sports, geo-, environmental protection, and industrial textiles. The transition was marked by newly discovered synthetic materials. Synthetics were easy to customize per requirements, with durability, strength, longevity, and low density compared to traditional textile materials.

Having achieved the summit in research and innovation in the area of technical textiles by incorporating the use of synthetics, researchers are looking towards sustainable materials to achieve the same feat. The green products drive started in 1987 with the United Nations release of the sustainable development concept. By the year 2004, enterprises started environmental compliance in the production process. The ASEAN community vision 2025 and the UN 2030 Agenda for sustainable development are taking the initiative forward by reviewing the progress for a complete sustainable chain from green sourcing and manufacturing to green marketing.

Green technology using natural biomaterials and anthropogenic fibers from natural sources to create high-value technical fibers, finishes, and products is presented in the book. Descriptions of common synthetic fibers, presently used in the technical textile sector, in terms of manufacturing technology, are also shared in the book. The book presents various aspects of cellulosic, anthropogenic cellulosic, and synthetic fibers and fundamentals, manufacturing, operations, classification, properties, and commercial applications. Special-technique fibers such as hollow fibers, bi-component fibers, melt-blown fibers, dry jet wet fibers, ultra-fine nanofibers, and gel spinning technology are presented. A brief synopsis of specialty fibers; frontier fibers; special commercial fibers; and their process, technology, and application is highlighted. Technologies for manufacturing all these superior new-concept fibers to manufacture high-value fabrics are addressed. Special weaving techniques for manufacturing high-tech structures that provide higher fabric strength, flexibility, toughness, and useable life are explained in the book. Attractive features of nanofibers for producing lower-weight functional fabrics with increased comfort and functionality are covered in the book. Agro-residue fibers from discarded parts of plants are included for their usefulness in agrotextiles, biocomposites, biosorption, and nanocellulose fabrication. The applicability of agro-residue is explored for technical textile finishes, and ecofriendly application technology is included. A case study series presented at the ends of the chapters reflect diverse fields, including gel spinning of polyethylene, used tire cords for outdoor furniture and installations, bird's nest fibers, electrospun chitosan-coated nanofibrous webs, peanut testa, chickpea testa, and onion peel functional finishes and color.

The field of technical textiles has been enriched by major advancements made in the last few decades. We have gathered the research of international academicians and technologists through their publications. The book aims to cover major developments and future directions in the field of technical textiles to achieve social and technological innovation. The book should serve as an important reference for students, technical textile professionals, academicians, and industry.

We wish to sincerely thank CRC Press for providing potential support and unstinting faith to us for the publication of this book.

We are very grateful to the Ministry of Textiles, Government of India, and Chandra Shekhar Azad University of Agriculture and Technology, Kanpur, India, for providing us all the support for conceptualizing and producing the entire book. We would like to thank our family members, colleagues, and all well-wishers for standing with us all the time. Finally, our thanks to Almighty God for directing us with his omnipresent power and positive force.

Authors
Dr. Mukesh Kumar Sinha
Dr. Ritu Pandey

About the Authors

Dr. Mukesh Kumar Sinha has been Joint Director (Research & Innovation), National Technical Textile Mission on deputation. The said project is on the growth of technical textiles in India, initiated by Ministry of Textiles, Government of India. Prior to that, he joined the Indian Defence Research and Development Organization (DRDO), Kanpur, in the year 2008, worked as Scientist F (Joint Director), and headed the division Textile Materials Research & Development (non-metallic materials for developing defensive protective textiles) and technical textile products. After obtaining a bachelor's in textile engineering (Shivaji University), master's (IIT Delhi), and PhD in textiles from the reputed Heriot-Watt University (formerly known as the Scottish College of Textiles, Scotland, UK) and also gained textile industrial experience abroad. He was awarded charted colorist from the Society of Dyers and Colorists (SDC), UK; Fellow of Institute of Engineers 2014 (FIE); Fellow Institute of Textile Association (2019); and DRDO Laboratory Scientist 2013 for his commendable work and efforts in the area of product-oriented research work for developing various defensive protective textiles (DPT) for Indian services. He has been a member of various textile technical organizations and represents various expert panels in the government of India. His technological credential research on functional technical protective textiles is reflected by 50+ publications in international peer-reviewed textile journals/international/ national conferences (invited talks), as well as two Indian patents filed in the technical textile thrust research area. He has also authored several book chapters (international publishers) in the context of technical textiles and functional fibers. He has represented/attended various textile conferences as expert/invited speaker and chaired various technical sessions.

Dr Ritu Pandey is Assistant Professor at Chandra Shekhar Azad University of Agriculture & Technology (CSAUAT) Kanpur, India. She has 27 years of teaching experience. Her alma mater is the prestigious Govind Ballabh Pant University of Agriculture & Technology (GBPUAT), Pantnagar, Uttarakhand, where she completed her bachelor's and postgraduate studies in textiles and clothing. She has guided several dozen postgraduate and PhD students so far. She has published more than 50 research papers and contributed book chapters. She has co-edited two books on museum textiles and co-authored one book on flax fiber processing technology. Her area of interest is plant fibers, dyes, and historic textiles. She has completed a research project on flax fiber as principal investigator. She was awarded a patent on accelerated retting of flax (gel retting) and filed another patent on lotus fiber drawing machinery. She has received the award of Outstanding Researcher for the last five years from her university for publishing research papers in the highest–impact factor journals. She has presented her research on plant fibers and dyes at various national and international forums, symposia, and conferences. She has been invited as a panelist to several conferences, including the UNESCO-sponsored International Forum of Natural Dyes and Fiber Technology held at Nantou, Taiwan.

1 Introduction— A Fundamental Overview

1.1 INTRODUCTION

Technical textiles are defined as textile materials manufactured with enhanced performance characteristics to improve their serviceability. Non-apparel uses of textiles fall under technical textiles. The technical textile industry serves humanity by catering to the needs of several industries, from agrotextiles to automobile, construction, environment, and medicine. Technical textiles is an emerging field that offers materials and superior products mainly in functional and technical performance rather than having decorative or aesthetic elements. Technical textiles are value-added functional textiles with widespread applications in several sectors for their innovative technical properties and high-performance structural characteristics. Technical textiles provide better comfort, safety, and performance standards and therefore are used in a variety of ways, ensuring enhanced lifestyle goals. Ever-increasing inventions and innovations in the field of textile processing are solving the problems of common consumers as well as limiting the disposal of hazardous effluents in the environment. Such innovations and futuristic designs are also adding value to technical textile segments. Technical textiles have gained popularity in several areas: agriculture, construction, environment, medical, and technical clothing components. Technical textiles in industrial segments also cater to the needs of protech, packtech, geotechnical, meditech, clothtech, buildtech, oekotech, and home tech [1]. There is overlap in all areas of technical textiles; for example, innovations in industrial textiles (indutech) are applicable in transport, home textiles, environment, medical, agro-geothermal, and sports clothing (Figure 1.1). Awareness regarding technical aspects was realized in the 1980s, but its commercialization and popularity started in the 1990s [2]. Presently the industry is more focused on development and innovation in technical textile segments and producing thousands of products according to demands in the relevant fields. The present scenario of ever-changing needs and demands calls for continuous research and development to provide the best to society.

This chapter aims to provide fundamental insights into technological developments that have helped to overcome the challenges faced by textile industries, especially high-tech technical textiles in producing lighter fabrics with optimized functional properties. This chapter is concerned mainly with the fundamental aspects of high-tech nonmetallic fibers and fabrics for textile industries. The book

DOI: 10.1201/9781003317074-1

Agrotech
Milk filtration
Protection nets
Mulch mats

Clothtech
Garment Interlining
Shoe components
Sewing threads
Insulation

Buildtech
Tarpaulin
Swimming pool liners
Fiber reinforced panel
Elevator belt
Pressure hose
Sun blinds
Composite

Mobitech
Car interior
Heat resistant seat covers
Car panels
Seat belt
Air bag

Oekotech
Dust filtration
Toxic waste sealing
Erosion protection
Air filter, oil filter

Indutech
Electrical cables
Jacquard harness
Conveyer belt
Optical fibers

Geotech
Check dam lining
Railway track stabilisation
Artificial turf
Reservoir lining
Waste pit lining

Protech
Electrostatic shielding
Muscle support
Heat resistant textile
Parachute, Sail cloth
Fire blanket

Meditech
Bypass filters
Ligaments, tendon, implants
Dialysis infusion
Prosthesis

Hometech
FR Carpet
Home furnishing
FR upholstery
Fiber fill mattress
Nonwoven

Packtech
Packing sack
Paper, composite
Hose, ropes
Jute bags

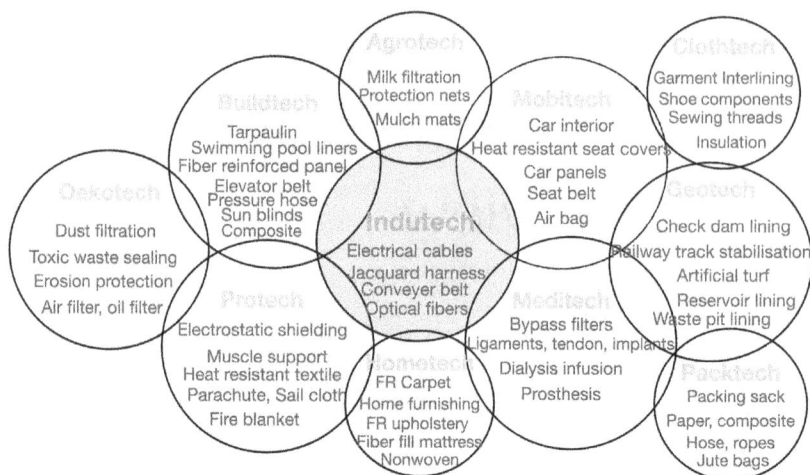

FIGURE 1.1   Technical textile segments and various products.

variously discusses the use of natural fibers, melt-blown fibers, spinning (non-woven) technology, multifunctional fibers/fabrics, functional auxiliaries, and finishing agents, as well as phase-change materials (PCMs) and specialized fabric weaving technology for specific end uses [3, 4].

The textile industry can be broadly classified into two categories: normal community textiles and technical textiles. In the case of normal community textiles, the technology used to produce fibers, yarns, and fabrics is quite generic, and the focus is mostly on aesthetic aspects rather than functional properties. To address this, technical textiles have been developed to manufacture fabrics with enhanced functional properties that can benefit various segments of the industry, namely agrotextiles (agrotech), building construction (buildtech), clothing (clothtech), ground engineering (geotech), home textiles (hometech), industrial textiles (indutech), medical textiles (meditech), automobile textiles (mobiltech), packaging textiles (packtech), protective textiles (protech), and sports textiles (sportech) [5]. Technical products made under each segment of technical textiles are presented in Table 1.1.

The technical textile industry is currently at a tipping point, moving from low-value products to high-end products. There is a vast scope on the global horizon to improve the consumption and production of technical textiles and turn it into an escalating growth sector. Technical textiles are distinguished by their functional properties and are engineered to fulfill specific purposes, often making them multifunctional. The development of these specialized products is heavily reliant on research and development efforts. However, the demand for high-tech technical textiles is on the rise globally, requiring advanced technical expertise in both product and process innovation. For instance, photoactive technical textiles can be utilized to create flexible solar panels, while spacer and auxetic fabrics find applications in various fields like home textiles, automotive textiles, medical,

TABLE 1.1

Utilization of Textile Fibers in Different Segments of Technical Textiles

Fiber Name	Application Sector
Kevlar (meta aramids)	Defensive protective armor textiles
Nomex para aramids	Automotive textile sectors
FR modacrylic	Firefighter suits
Ultra-high-molecular-weight polyethylene	Stab and ballistic suits
Superabsorbent fibers	Moisture management fabrics (sportswear)
Carbon fibers	Aerospace industries and lighter-weight composites
Glass fibers	High-strength lighter-weight and durable composites (architecture textiles)
Flame-retardant (FR) viscose	Industrial safety gears
Anti-microbial/anti-fungal/anti-bacterial fibers	Medical textiles
Chameleon fibers	Camouflaging textiles
Polyphenylene sulfide (PPS) fibers	
Polytetrafluoroethylene (PTFE)	Inert and chemical-resistant textiles
Polybenzimidazole fibers (PBI)	Firefighter gear, astronaut spacesuits
Polybenzoxazole (PBO)	Sports textiles
Phenolic fibers and bead materials	Defensive protective toxic chemical warfare suits
Conductive fibers	Smart textiles
Activated carbon fabrics	Toxic warfare agents, chemical adsorbent textiles, and composites
Biodegradable fibers	Medical textiles and substitutes for plastics and metals with suitable technical textiles materials, particularly in those areas where it is possible to reduce environmental pollution by using biodegradable technical textile products

security, and aviation [6]. Addressing technological gaps in three key areas is critical for the global technical textile industry:

- Technology of high-performance/functional fibers (FFs)
- Specialized coating and lamination (SCL)/finishing processes
- Specialized manufacturing technologies such as triaxial weave and leno weave

1.2 TECHNOLOGY OF HIGH-PERFORMANCE TECHNICAL FIBERS

High-performance (HP) fibers with high tenacity, modulus, and resiliency; resistance to wear, fire, and chemicals; and dimensional stability are superior to normal commodity fibers. High-performance fibers have an amalgamation of properties to fulfil the demand of technical textile segments. Some commercially available high-performance fibers include silicon carbide (Nicalon), p-aramid (Kevlar), poly p-phenylene benzobisoxazole (PBO Zylon), and ultra-high-molecular-weight polyethylene (UHMWPE spectra and Dyneema). HP fibers are made to suit unique specific functional requirements. These properties are higher tensile strength and chemical and thermal resistance as compared to the normal grade of polymeric fibers. Polymeric high-performance fibers have a tenacity range of ~15–50 gpd and tensile modulus of ~250–1500 gpd, which is approximately 20 times greater than normal grade of polymeric fibers. Functional fibers, on the other hand, have specific applications, such as moisture management and antimicrobial, conducting, and bacterial antistatic properties. Multinational global companies like DuPont, Honeywell, DSM, and Nippon have global trade monopolies for manufacturing such high-performance and functional fibers [7–9]. However, there is a need for more global players to develop indigenous versions of these special fibers to avoid dependence on imports and become self-reliant in this area, especially in third world countries. Some important high-performance fibers (HPFs) and functional fibers are presented in Table 1.1.

Fiber spinning, even for common types of fibers, is a complex process that involves using specialized machines and spinnable grades of polymers. Different techniques are employed for fiber spinning, such as wet-dry spinning, wet solution spinning, dry jet wet spinning, and melt spinning. Once the fibers are spun, they undergo a drawing process to enhance their strength and crystallinity, which enables them to possess particular functional characteristics. This entire process and technology can be quite intricate and requires significant attention. Numerous universities and public-funded research institutions are engaged in researching high-performance and functional fibers in the textile industry. However, private-sector corporations such as DuPont have made significant contributions to R&D. One of the key focus areas of research for regenerated fibers is to develop environmentally friendly manufacturing technology and fibers with specific characteristics for targeted end uses. Notable R&D centers in this area include Lenzing in Austria and Birla Viscose in India. In addition, synthetic fibers are being developed by both private and public organizations across the US, Europe, and Asia. Current research in this field are focused on developing fibers with enhanced properties, as cited in the following.

The Institute fur Textiltechnik in Aachen, Germany, has conducted research into the development of bicomponent fibers using polyphenylene sulfide, a high-performance polymer, in combination with polyethylene terephthalate (PET), a standard polymer. These fibers have potential applications in various fields, such as drying filters and geotextiles, due to the excellent chemical resistance and high temperature stability of PPS. As PPS has not been previously used in bicomponent fibers, there is a lack of experience in melt spinning techniques for these materials.

A team of researchers from Advanced Fibres, Switzerland, have successfully developed HP bicomponent fibers called polyphenylene sulfide in combination with polyethylene terephthalate through a process called melt spinning. This innovative approach enabled the team to use PPS as either a core or a sheath material, resulting in specialized functionalities such as enhanced thermo-bonding potential, improved flame, and chemical resistance. The research team also explored the parameters required to ensure processing of both PPT and PPS during coaxial extension with varying core sheath volume ratios. This investigation facilitated the optimization of the manufacturing process and allowed for the creation of high-quality bicomponent fibers with specific properties tailored to the desired application [10].

European textile industries are investing in research and development to produce high-performance textile products such as bio and nanofibers to ensure their competitiveness globally. Dow Fibre Solutions has recently created a new fiber called XLA, based on olefin, which maintains its stretch properties even when exposed to high temperatures of up to 220 °C. Meanwhile, DuPont's Bio-PDO polysaccharide fibers are made from renewable resources like corn.

Polytetrafluoroethylene is a synthetic fluoropolymer with a wide range of applications. Among these, Teflon-coated fiberglass fabric is one of the most durable materials used in architecture. Originally used as a roofing material in 1973, it has since become a popular choice for lightweight structures due to its longevity and strength.

Netherlands-based company DSM Dyneema has developed a new grade of gel-spun polyethylene fiber called Dyneema Purity. This high-performance, ultra-strong fiber is designed for use in biomedical applications, particularly for implants used in medical surgery. The use of this fiber can potentially reduce scarring and shorten recovery times for patients [11].

A new high-performance polyetherimide (PEI) fiber designed for transportation applications has been developed by a research team at GE Plastics. This development is significant, as aramid fibers have become scarce in supply, and polyetherimide fiber has emerged as a sought-after material for transportation and military applications due to its inherent flame-retardant properties and exceptional performance.

Honeywell has developed a new fiber called Spectra 53000 using a patented gel spinning method made from ultra-high-molecular-weight polyethylene using a patented gel-spinning process. This fiber is being utilized in ballistic applications due to its exceptional performance. The company claims that Spectra 53000 provides up to 20% better performance than spectra products in ballistic uses [12]. DuPont, based in Wilmington, Delaware, has recently created a new variant of Kevlar fiber called Kevlar QV that has a 300-denier count and is specifically designed for use in body armor worn by officers. The company claims that this fiber technology offers exceptional resistance against punctures and effectively hinders sharp objects (stab resistant) from penetrating the armor by dissipating energy and preventing the Kevlar fibers from being fractured. Furthermore, when ultra-high-molecular-weight polyethylene (UHMWPE) is combined with other Kevlar hybrid ballistic materials, this new fibrous laminate can also provide protection against threats from bullets and handmade weapons. The fibrous laminated vests act as defense armor protective textiles.

Zoltek Corp., a company based in St. Louis, specializes in manufacturing carbon fibers that are used for various structural or industrial applications. These applications require fibers with exceptional properties such as high strength and stiffness, depending on the intended use [13]. To cater to the demand for flame-retardant fabrics, Zoltek Corp. has developed a fiber called Pyron. Pyron is an oxidized polyacrylonitrile fiber that offers the high-temperature and flame-resistant capabilities of carbon fibers while also being easily manageable for textile operations. Pyron is inherently resistant to fire, making it an ideal material for flame-retardant fabrics.

Magellan Systems International, located in Arnhem (NL), has recently developed a new synthetic fiber called M5/polyhydroquinone-diimidazopyridine (PIPD). This fiber claims exceptional strength, surpassing that of para-aramid and carbon fibers, as well as having a higher Young's modulus than PBO fibers. M5 is made from a polymer based on 2,6-dihydroxy-terephthalic acid and 2,3,5,6-tetra-aminopyrimidine. The company claims that M5's mechanical properties make it a strong competitor in carbon fiber applications, with additional benefits such as easy fabrication into composite form, the ability to absorb high levels of energy at the time of damage, and high electrical resistance. Furthermore, M5 exhibits high damage tolerance, making it an ideal choice for various applications [14].

Nexia Biotechnologies, based in the USA, has developed a new synthetic fiber called BioSteel made from recombinant spider silk. This product is expected to have various applications in medical, military, and industrial markets due to its eco-friendly nature, biodegradability in water, and non-polluting production process [15]. The BioSteel material, produced by genetically modifying goats with spider genes, boasts exceptional strength, being 20 times stronger than steel, with a breaking strength of around 300,000 pounds per square inch. It is also lighter than synthetic, petroleum-based polymers by 25%, making it suitable for applications that require strength and lightness, such as bulletproof clothing, racing vehicles, and aircraft. Additionally, BioSteel has potential uses in medical joints and fishing lines. Overall, BioSteel has the potential to revolutionize various technical textile industries.

1.2.1 FIRE-RETARDANT CLOTHING

A team of researchers from the Ecole Nationale Superiere des Arts et Materialux in Roubaix have been conducting a study to develop high-performance fibers that are both lighter and safer. As part of their research, they compared the reactions of different high-performance fibers and found that PBO fibers outperformed p-aramid fibers in terms of heat generation. Furthermore, the researchers noted that PBO fibers did not contribute to the spread of fire and generated less smoke compared to p-aramid fibers [16].

1.2.2 COMPOSITE APPLICATIONS

Advanced composites made from high-performance fibers, including materials such as high-modulus polyethylene (PE), boron, quartz, ceramic, poly p-pbenylene-2, 6-benzobisoxazole, and hybrid combinations, are increasingly being utilized in a wide range of industries. These industries include boat building, automotive, sports, aerospace, fuel, transportation, infrastructure, anti-corrosive applications, unmanned aerial vehicles (UAVs) and unmanned combat aerial vehicles (UCAVs) [17, 18].

1.3 SPECIALIZED COATING AND LAMINATION/ FINISHING PROCESSES

The process of coating involves the application of a layer of a substance, typically a polymer, onto the surface of a fabric to enhance its properties such as durability, water resistance, and color fastness. This layer can be added through several methods, including spraying, roll coating, or dipping. Coated fabrics are commonly used in outdoor gear, tents, and clothing. On the other hand, laminating involves the bonding of two or more layers of materials to create a composite material with specific properties. For instance, a foam layer can be laminated to a fabric to create a cushioning effect, or a film layer can be laminated to a fabric to make it waterproof. Laminating is frequently employed in industrial, medical, and automotive applications to create high-performance materials. The textile industry is yet to fully utilize intriguing and innovative "smart" materials, like phase-change materials, temperature and shape memory polymers, and surface modification techniques that enhance adhesion [18–20]. In conclusion, combining the processes of coating and laminating offers a broad range of possibilities for improving and modifying fabrics, as well as producing innovative products with unique characteristics, which will be discussed further in Chapter 3.

1.4 SPECIALIZED MANUFACTURING TECHNOLOGIES SUCH AS TRIAXIAL WEAVE AND LENO WEAVE

Triaxial weave has three sets of yarns; warp, weft, and a bias stuffer. Yarns interlace at 60° with each other. The leno weave facilitates the twisting of adjacent warp yarns together, enabling them to cross over each other and thus trap the interlacing weft yarns. This holds the yarns firmly, making the fabric strong and suitable for technical applications in medicine, construction, and agriculture [21]. Descriptions of specialized weave manufacturing technologies are covered in Chapter 3.

1.5 CONCLUSION

Technical textiles are value-added functional textiles with wide range of applications in different sectors for their innovative technical properties and high-performance structural characteristics. Technical textiles provide better comfort, safety, and performance standards and are therefore used in a variety of ways, ensuring enhanced lifestyle goals. Technical textiles are manufactured with fibers containing inherent functional properties. Fabrics made of conventional or technically superior fibers/yarns are improved further with the high-tech finishes required for the intended use. Important roles are played by technical fabrics in the fields of agriculture; building construction; biomedical; infrastructure; and some specialized apparel for sports, skiing, and hiking activities. Ever-increasing inventions and innovations in the field of textile processing are solving the problems of common consumers as well as limiting the disposal of hazardous effluents in the environment. Such innovations and futuristic designs are also adding value to the technical textile segments. Hence, the technical textile sector is constantly becoming bigger and benefiting humankind.

REFERENCES

1. Chowdhury A, Dhamija S. 2021. Manufacturing technologies and scope of advanced fibres. In Recent Trends in Traditional and Technical Textiles: Select Proceedings of ICETT 2019. Springer, Singapore, 179–187. https://doi.org/10.1007/978-981-15-9995-8_16
2. Horrocks AR, Anand SC. 2000. Handbook of Technical Textiles. Elsevier, Cambridge.
3. Sarier N, Onder E. 2012. Organic phase change materials and their textile applications: An overview. Thermochimica Acta 540:7–60. https://doi.org/10.1016/j.tca.2012.04.013
4. Iqbal K, Khan A, Sun D, Ashraf M, Rehman A, Safdar F, Basit A, Maqsood HS. 2019. Phase change materials, their synthesis and application in textiles—A review. The Journal of the Textile Institute 110(4):625–638. https://doi.org/10.1080/00405000.2018.1548088
5. Belso-Martínez JA, Tomás-Miquel JV, Expósito-Langa M, Mateu-García R. 2019. Delving into the technical textile phenomenon: Networking strategies and innovation in mature clusters. The Journal of the Textile Institute 111(2):260–272. https://doi.org/10.1080/00405000.2019.1631638
6. Aldalbahi A, El-Naggar ME, El-Newehy MH, Rahaman M, Hatshan MR, Khattab TA. 2021. Effects of technical textiles and synthetic nanofibers on environmental pollution. Polymers 13(1):155. https://doi.org/10.3390/polym13010155
7. Fei B. 2018. High-performance fibers for textiles. In Engineering of High-Performance Textiles. Woodhead Publishing, 27–58. https://doi.org/10.1016/B978-0-08-101273-4.00002-0
8. Sinha MK, Das BR, Prasad N, Kishore B, Kumar K. 2018. Exploration of nanofibrous coated webs for chemical and biological protection. Zaštita Materijala 59(2):189–198. https://doi.org/10.5937/ZasMat1802189K
9. Zhang Q, Li X, Dong J, Zhao X. 2022. High-performance polyimide fibers. Advanced Industrial and Engineering Polymer Research 5(2):107–116. https://doi.org/10.1016/j.aiepr.2022.03.004
10. Perret E, Reifler FA, Hufenus R, Bunk O, Heuberger M. 2013. Modified crystallization in PET/PPS bicomponent fibers revealed by small-angle and wide-angle X-ray scattering. Macromolecules 46(2):440–448. https://doi.org/10.1021/ma3021213
11. Heisserer U, Van der Werff H. 2012. The relation between Dyneema® fiber properties and ballistic protection performance of its fiber composites. In 15th International Conference on Deformation, Yield and Fracture of Polymers 3(3):242–246.
12. Chen X, Zhou Y. 2016. Technical textiles for ballistic protection. In Handbook of Technical Textiles. Woodhead Publishing, 169–192. https://doi.org/10.1016/B978-1-78242-465-9.00006-9

13. Newcomb BA. 2016. Processing, structure, and properties of carbon fibers. Composites Part A: Applied Science and Manufacturing 91:262–282. https://doi.org/10.1016/j.compositesa.2016.10.018

14. Afshari M, Sikkema DJ, Lee K, Bogle M. 2008. High performance fibers based on rigid and flexible polymers. Polymer Reviews 48(2):230–274. https://doi.org/10.1080/15583720802020129

15. Koumoulos EP, Trompeta AF, Santos RM, Martins M, Santos CMD, Iglesias V, Böhm R, Gong G, Chiminelli A, Verpoest I, Kiekens P. 2019. Research and development in carbon fibers and advanced high-performance composites supply chain in Europe: A roadmap for challenges and the industrial uptake. Journal of Composites Science 3(3):86. https://doi.org/10.3390/jcs3030086

16. Dhineshbabu NR, Bose S. 2019. UV resistant and fire retardant properties in fabrics coated with polymer based nanocomposites derived from sustainable and natural resources for protective clothing application. Composites Part B: Engineering 172:555–563. https://doi.org/10.1016/j.compositesb.2019.05.013

17. Esthappan SK, Sinha MK, Katiyar P, Srivastav A, Joseph R. 2013. Polypropylene/zinc oxide nanocomposite fibers: Morphology and thermal analysis. Journal of Polymer Materials 30(1):79–89.

18. Chavan S, Kanu NJ, Shendokar S, Narkhede B, Sinha MK, Gupta E, Singh GK, Vates UK. 2023. An insight into nylon 6, 6 nanofibers interleaved E-glass fiber reinforced epoxy composites. Journal of The Institution of Engineers (India): Series C 104(1):15–44. https://doi.org/10.1007/s40032-022-00882-0

19. Sinha MK, Das BR, Srivastava A, Saxena AK. 2013. Needleless electrospinning and coating of poly vinyl alcohol with cross-linking agent via in-situ technique. International Journal of Textile and Fashion Technology 3(5):29–38.

20. Shim E. 2019. Coating and laminating processes and techniques for textiles. In Smart Textile Coatings and Laminates. Woodhead Publishing, 11–45. https://doi.org/10.1016/B978-0-08-102428-7.00002-X

21. Bilisik K. 2012. Multiaxis three-dimensional weaving for composites: A review. Textile Research Journal 82(7):725–743. https://doi.org/10.1177/0040517511435013

2 Emerging Technical Fiber Technology

2.1 INTRODUCTION

This chapter delves into the exciting world of futuristic fibers for the high-tech technical industry. The aim of this chapter is to present various aspects of fiber technology that can help textile conservators obtain sustainable, biodegradable, eco-friendly, and cost-effective technical textile fabrics. In the 21st century, innovative technical and high-performance fibers and textiles could even possess anti-biological functions. Today's world requires technical textile fabrics that are sustainable for the environment and have a lifespan of over 100 years. Researchers worldwide are striving to develop improved textile materials to secure the future, with a particular focus on spinnable polymeric materials that offer light weight, stretchability, ease of dyeing, optical transparency, chemical and biological resistance, higher strength and modulus, and most importantly photo stability [1–3]. The material selection, type of weave, and fabric construction will be decided based on meeting the current requirements of technical textile fabrics. Currently, the technical textile industry mostly relies on using normal commodity synthetic and natural fibers to manufacture various high-end technical textile products. However, these fibers have limitations in terms of stability, strength/modulus, and functionality to perform per technical textile applications and their stringent quality control norms. Commodity fibers are prone to damage from light, environmental factors, fungal and bacterial degradation, and applied dye/finishes and have low fastness to light, washing, rubbing, and abrasion.

To address these issues, innovative new applications of specialty fibers should be developed and explored for innovative technical textile applications. It is important to widen the selection of fibers from normal commodity to substitute and explore their technical applications. The technical textile industry has not paid much attention to developing frontier fibers, functional fibers, and high-performance fibers and subsequently their use in making high-performance technical textile fabrics. However, significant research is being done on the development of new fibers for technical applications worldwide. These new developments have been made possible through modifications of fiber surfaces. It is time to switch the technical textile industry from normal commodity fibers to the technical textile fibers to fill the technological gap. Therefore, technical textiles should tap into the gray areas of fiber technology, particularly specialty innovative fibers [4].

DOI: 10.1201/9781003317074-2

Abbreviations used in the chapter	
Abbreviations	**Full form**
CMC	Ceramic matrix composite
FEP	Fluorinated ethylene polymers
HDPE	High density polyethylene
HM-HT	High-modulus-high-tenacity
HMW	High modulus weight
HMPE	High modulus polyethylene
HP	High performance
MMC	Metal matrix composite
PBI	Polybenzimidazole
PBO	P-phenylene-2,6-benzobisoxazole
PCS	Polycarbosilane
PE	Polyethylene
PET	Polyethylene terephthalate
PEEK	Polyethyletherketone
PLA	Polylactic acid
PP	Polypropylene
PPS	Polyphenylene sulfide
PTFE	Polytetrafluoroethylene
PVDF	Poly(vinylidene fluoride)
PVF	Poly(vinyl fluoride)
UHMWPE	Ultra-high-molecular-weight polyethylene
UV	Ultraviolet

2.2 CLASSIFICATION OF FIBERS

Specialty fibers of technical grades and usages refer to thread-like structures that are thin, long, and flexible, with a high length to width ratio of at least 100:1. These fibers can be obtained from natural sources or can be human made and are typically spun into yarn before being woven or knitted, or fibers are directly converted onto nonwoven structures [5]. In addition to natural and human-made fibers, specialty fibers include those with unique properties for specific technical applications (Table 2.1). Examples and classifications of these fibers are described in Figure 2.1.

2.2.1 NATURAL FIBERS

Natural fibers refers to hair-like materials that are obtained directly from plants and animals. These fibers possess certain characteristics such as the ability to be spun into yarn, durability, strength, high moisture absorption capacity, and a pleasant appearance and texture. Additionally, natural fibers can be categorized into three subgroups based on their origin: plant fibers, animal fibers, and mineral fibers [6–9].

TABLE 2.1

Physical Properties of High-Performance Fibers

Fibers	Moisture Regain (%)	Elongation (%)	Tenacity (g/d)	Specific Gravity	Limiting Oxygen Index
Aramid 3 (*meta*)	6.5	22–32	4.0–5.3	1.38	30
Aramid (*para*)	4.0	2.5–4.0	21–27	1.44	29
Glass	<0.10	4.8	15.3	2.5	>100
HDPE 4	0	2.8–4.4	24–35	0.97	-
Melamine	5	18	1.8	1.44	32
Novaloid phenolic	6.0–7.3	30–60	1.3–2.4	1.27	33
PAN/carbon	9	19	24	1.4	55
PBI	15	25–30	2.6–3.0	1.43	38
PBO	0.6–2.0	2.5–3.5	42	1.5	68
Poly-acrylate	12	20–30	1.3–1.7	1.5	43
Poly-acrylate liquid crystal	<0.10	3.8	26–29	1.41	37
Poly-amide Imide	4–4.5	16–21	3.4	1.34	32
Poly-Imide	3	19–21	3.7	1.41	40
PPS	0.6	40	3.5	1.37	34
PTFE	0	19–140	0.9–2.0	2.1	>100

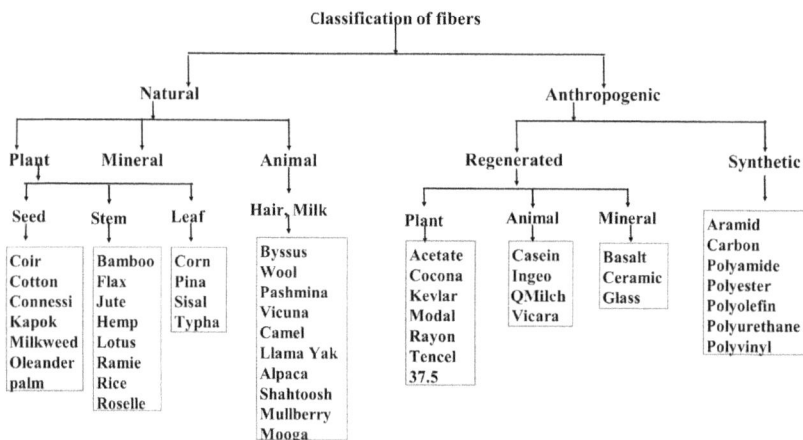

FIGURE 2.1 Classification of fibers.

2.2.2 HUMAN-MADE FIBERS

Human-made fibers are synthetic fibers that are created by humans through the use of various chemicals in industrial processes. These fibers are made from polymers, which are large molecules made by combining smaller repeating units.

Some well-known examples of human-made fibers include rayon, nylon, polyester, acrylic, and acetate. Human-made fibers can be classified into three categories: regenerated fibers, synthetic fibers, and inorganic fibers. Regenerated fibers are produced by using natural materials like cellulose, while synthetic fibers are created entirely from chemicals. Inorganic fibers, on the other hand, are made from materials like glass or metal.

2.2.2.1 Regenerated Fibers

These fibers are also known as semi-synthetic fibers. Cellulose obtained from plants is purified, and then fibers are produced from it. These are made of long-chain polymers, which are modified by a chemical process to enable polymerization to form fibers, for example, viscose, rayon, and bamboo.

2.2.2.2 Synthetic Fibers

These fibers are formed by the polymerization of monomers. Once a polymer is formed, it is converted into a fluid form. The dissolved or molten polymer is extruded through narrow holes to give filaments, for example, polyester, acrylic, and nylon.

2.2.2.3 Inorganic Fibers

These are also known as metallic fibers. They are obtained from copper, silver, or gold and can be extruded from nickel or iron. Glass fibers, micro glass fibers, carbon fibers, activated carbon fibers, rock wool fibers, meerschaum (hydrated magnesium silicate), ceramic fibers, potassium titanate fibers, and Wollastonite (calcium silicate fiber) are some examples of inorganic fibers. These fibers have a higher melting point and are conductive.

2.2.3 SPECIALTY FIBERS

Specialty fibers can be further classified as presented in Figure 2.2 and Table 2.2.

Specialty fiber's content and manufacturing technology is designed for diverse applications. The fibers, stapled, or continuous tow, include any of the following:

1. High-performance fibers, organic fibers, and inorganic fibers
2. Inherent functional fibers
3. Biodegradable fibers

2.2.3.1 High-Performance Fibers

High-performance fibers are a type of material known for their exceptional mechanical properties, such as high strength; high modulus; and resistance to wear, deformation, chemicals, and high temperatures. They are widely used in the production of engineered textiles, industrial fabrics, and other high-performance textiles to provide solutions to a range of technical challenges, including environmental issues, personal safety, security, health, and comfort.

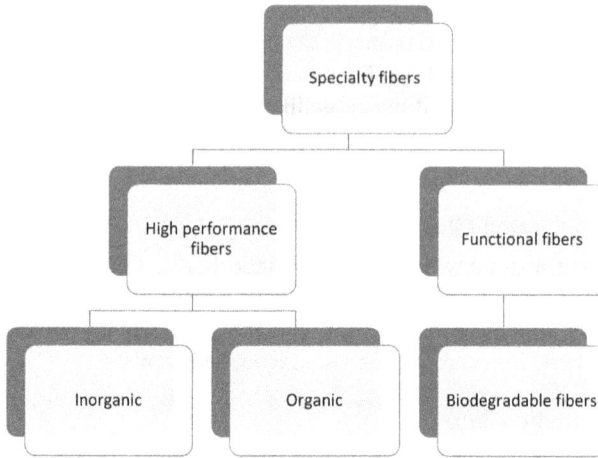

FIGURE 2.2   Sub-classification of specialty fibers.

TABLE 2.2
Categorization of Specialty Fibers

Fiber	Categorization*
Kevlar (meta aramid)	HPF
Nomex (para aramid)	HPF
Fire-retardant modacrylic	FF
Ultra-high-molecular-weight polyethylene	HPF
Superabsorbent fiber	FF
Carbon fiber	HPF
Glass fiber	HPF
Flame retardant (FR) viscose	FF
Anti-microbial/anti-fungal/anti-bacterial fiber	FF
Chamaeleon fiber	FF
Polyphenylene sulfide fiber (PPS)	Both HF and FF
Polytetrafluoroethylene (PTFE)	FF
Polybenzimidazole fiber (PBI)	FF
Polybenzoxazole (PBO)	HPF
Phenolic fiber	HPF, FF
Conductive fiber	FF

*HPF—High-performance fibers, FF—Functional fibers

These fibers can be categorized into three main groups: polymeric, inorganic, and metallic fibers. Polymeric high-performance fibers have a tenacity of 15–50 g/denier, elongation to break of 0.5–15%, and modulus of elasticity of 20–4000 g/denier. Although the strength and modulus of non-polymeric fibers are also high, their density or specific strength and modulus are not as high as those of polymeric high-performance fibers [10].

Using high-performance fibers and finishes in the production of technical textile fabrics offers several advantages over commodity synthetic and natural fibers. These fibers are inherently functional and sustainable over a longer period of time, and their high strength and modulus make them ideal for creating durable fabrics. They are also easy to dye and finish and can firmly bond with dyestuff and finishes for long-lasting colorfastness [11].

High-performance fibers can pose challenges in the conversion process to woven textiles or nonwoven fabrics, as well as in the application of specialty coatings and finishes. Nonetheless, they remain an essential resource for meeting the technical demands of modern society and addressing a diverse range of challenges. In the manufacturing of technical textile fabrics, various types of fibers are utilized; each offering unique properties that contribute to achieving specific desired characteristics. For example, materials with heat- and flame-resistant qualities such as phenolic fibers, polybenzimidazole, and polytetrafluoroethylene are ideal for producing protective clothing, while high-modulus polyethylene is preferred for applications involving ballistic protection and rope production [4, 12]. Additionally, chemically stable polymers like polyphenylene sulfide and polyether ether ketone are commonly used in filtration and other chemically aggressive environments. Aramid fibers are a significant type of fiber, commercially classified into two categories: meta-aramid fibers (m-aramid) and para-aramid fibers (p-aramid). m-aramid fibers have good thermal stability, but their mechanical properties are poor due to the inability of their polymer chains to closely pack together. On the other hand, p-aramid fibers have closely aligned chains, resulting in much better and higher strength. A team of researchers from the Ecole Nationale Supérieure des Arts et Materiaux in Roubaix have been conducting a study to develop high-performance fibers that are both lighter and safer. As part of their research, they compared the reactions of different high-performance fibers and found that PBO fibers outperformed p-aramid fibers in terms of heat generation. Furthermore, the researchers noted that PBO fibers did not contribute to the spread of fire and generated less smoke compared to p-aramid fibers.

Table 2.1 summarizes the essential properties of various aramid fibers, with each type offering specific benefits to achieve the targeted properties of technical textile products. Polyphenylene sulfide is a high-performance polymer known for its excellent chemical resistance and high temperature stability, making it suitable for a wide range of applications such as filters for drying and geotextiles. While PPS has not been used in bi-components before, its potential as a core or sheath material for melt spinning with polyethylene terephthalate has been explored. The method used in this study presents advantages such as enhanced thermobonding capability, flame retardancy, and chemical resistance. Researchers also examined various parameters to ensure consistent processing of PPS and PET at different core sheath volume ratios. PPS exhibits favorable characteristics such as high extensibility, tear strength, elasticity, low moisture absorbency (which improves its anti-fungal properties), and increased abrasion resistance.

2.2.3.2 Inherent Functional Fibers

Functional fibers possess inherent qualities such as hydrophilicity, hydrophobicity, flame resistance, chemical resistance, antibacterial properties, and conductivity, among others. End-users select functional fibers based on their specific requirements. Compared to traditional coated fabrics, these fibers offer superior performance. When producing high-quality textile raw materials suitable for technical and non-technical applications, it is important to ensure they possess good biodegradability and compatibility and at least acceptable mechanical properties. Surface-related characteristics, including hydrophilicity, hydrophobicity, oil and soil repellence, water penetration resistance, and antibacterial properties, are also important considerations. Materials used in heritage and medical textiles are particularly susceptible to damage from biological agents, such as bacteria, fungi, and viruses, so it is vital to protect these materials. Technical fibers are available for specific applications, such as moisture absorption and antibacterial properties, electrical conductivity, anti-static properties, and anti-odor properties. Specialty technical fibers such as FR Modacrylic, superabsorbent fibers, flame-retardant viscose, flame-retardant polyester, high tenacity/super high tenacity nylon, high tenacity/super high tenacity polyester, high tenacity/super high tenacity polypropylene, high tenacity/super high tenacity viscose, and anti-microbial/anti-fungal/anti-bacterial fibers are widely used in various applications. To enhance the durability of technical textiles, research and development have focused on incorporating bactericidal agents into polymers. However, most antimicrobial properties are achieved through a slow-release model, which releases the biocidal agent to the material's surface, resulting in the inactivation of microorganisms. This method has limited durability of the biocidal property [4, 13].

2.2.3.3 Biodegradable Fibers

In response to increasing environmental concerns, there has been a recent surge in the development of biodegradable fibers within the textile industry [7]. These fibers are designed to decompose into harmless materials when disposed of, making them ecologically favorable. Biotechnology advancements have enabled the creation of natural organic fibers with superior properties, high yields, and value-added eco-friendly features. Furthermore, reinforcing the need for eco-friendly processes and products includes promoting research and innovative technology for developing sustainable textiles. Biodegradable fibers for high-tech technical textile applications maintain the eco-environment and a non-hazardous textile industry [14]. These critical/sustainable technical textile technology research areas should be identified for various end uses and application sectors that focus mainly on biodegradable technical textiles [15]:

- Development of biodegradable polymers
- Systems and processes for the recyclability aspect of technical textiles
- Innovative methodology for disposability of waste into recyclable technical textile materials

- Process and technology development for the conversion of non-biodegradable polymeric fibers into bio-degradable fibers for safeguarding the environment
- Research on making textiles more sustainable, like water consumption reduction, increasing the use of renewable energy
- Substitutes for plastics and metals with suitable technical textile materials, particularly in those areas where it is possible to reduce imports and develop eco-friendly technical textile products
- Good-quality wood pulp—Absorbent fiber material for hygiene products

By incorporating biocompatible and biodegradable fibers into operations, the textile industry can benefit from a cost-effective and sustainable approach to production. Key examples of biodegradable fibers include PLA, Tencel, viscose, cotton, and wood cellulose. There are various types of natural fibers, including banana fiber [16], flax [17–19], lotus [20], angora rabbit hair fiber, pineapple fiber [6], sisal, mesta, and silk [21], which are often referred to as "forgotten fibers" in the textile industry. The use of natural fibers has recently gained popularity, particularly in the technical textile fabric industries, due to the demand for eco-friendly and biodegradable products. Additionally, natural fibers offer several advantages over synthetic fibers. There are also new applications for natural fibers, such as flax, hemp, wool, silk, lotus, elephant [22], and mohair, including natural fiber–reinforced composite and blended materials for use in technical textile fabric industries. One highly beneficial example of this is starch-based biodegradable fibers used in technical apparel fabric. These fibers are drawn in a process that improves their mechanical properties, making them stronger and more durable. They can be used in the manufacturing of various articles, including those with open, semi-densely packed, or densely packed structures. Another segment of biodegradable fibers involves synthetic polymers mixed with natural fibers to create multi-component biodegradable fibers, similar to spider silk webs. Overall, the use of natural fibers promotes sustainability and environmental responsibility. In summary, we can conclude that in India, we are now moving forward in converting biodegradable and agrowaste textile materials into technical textile products. Such sustainable technical textile technology will be critical for the growth and development of futuristic technical textile applications, balancing the ecosystem and ensuring the country's environmental sustainability.

2.3 PROCESS AND TECHNOLOGY

Traditional methods for producing polymer fibers with exceptional mechanical properties involve techniques such as melt spinning, dry spinning, wet spinning, and gel-state spinning. These methods utilize mechanical forces to extrude a polymer melt or solution through a spinneret and subsequently draw the resulting filaments as they solidify or coagulate. Once solidified, the filaments can be combined to form threads and/or drawn to alter their properties [23, 24]. Figure 2.3 illustrates the spinning technology and process for various polymers.

```
                        ┌─────────────────┐
                        │ Organic polymers│
                        └─────────────────┘
            ┌───────────────────┴──────────────────────────┐
    ┌───────────────┐                            ┌──────────────────┐
    │Stiff molecules│                            │Flexible molecules│
    └───────────────┘                            └──────────────────┘
      ┌──────┴──────┐            ┌───────────────┬──────┴──────┬───────────────┐
 ┌────────┐  ┌────────┐     ┌────────┐     ┌────────┐     ┌────────┐
 │ Wet    │  │ Melt   │     │ Dry    │     │ Wet    │     │ Melt   │
 │spinning│  │spinning│     │spinning│     │spinning│     │spinning│
 └────────┘  └────────┘     └────────┘     └────────┘     └────────┘
                                                    ┌─────────┬──────────┐
                                               ┌─────────┐ ┌────────┐
                                               │ Super   │ │ Normal │
                                               │structure│ │spinning│
                                               └─────────┘ └────────┘

  Aramids    Aromatic       Cellulose      UHMW      HMW       Nylon
             polyesters      acetate         PE        PE       PP PE
```

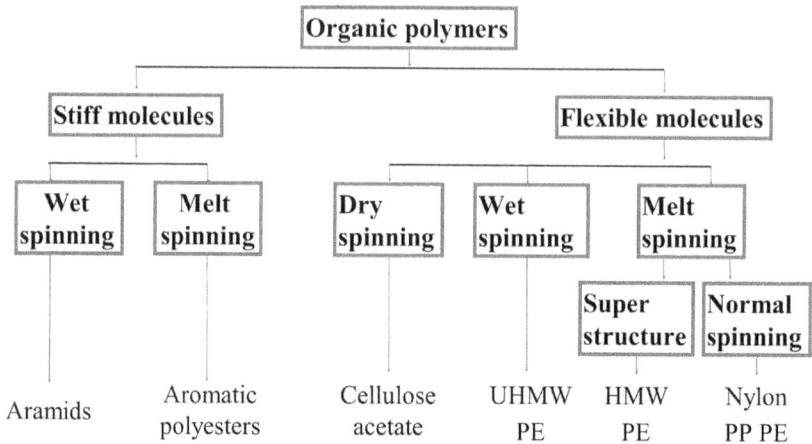

FIGURE 2.3   Schematic diagram for fiber spinning technology and process.

TABLE 2.3

Various High-Performance Fibers and Their Manufacturing Technology

Fibers	Technology
Meta aramids	Melt spinning
Para aramids	Dry-jet-wet spinning
Carbon fibers	Pitch melt spinning
Ultra-high-molecular-weight polyethylene	Gel spinning
PBO (p-phenylene-2,6-benzobisoxazole)	Jet-wet spinning
Flame-retardant viscose rayon	Wet spinning

Following are the spinning processes and technology used for human-made fiber production:

1. Dry spinning
2. Wet spinning
3. Melt spinning
4. Dry-jet spinning
5. Gel spinning

Table 2.3 provides a summary of the technology and processes used to produce various high-performance and functional fibers.

2.3.1 MELT SPINNING

Melt spinning is a commonly employed technique for fiber spinning in which either polymer pellets are melted or molten polymer is used. The extruded filaments are

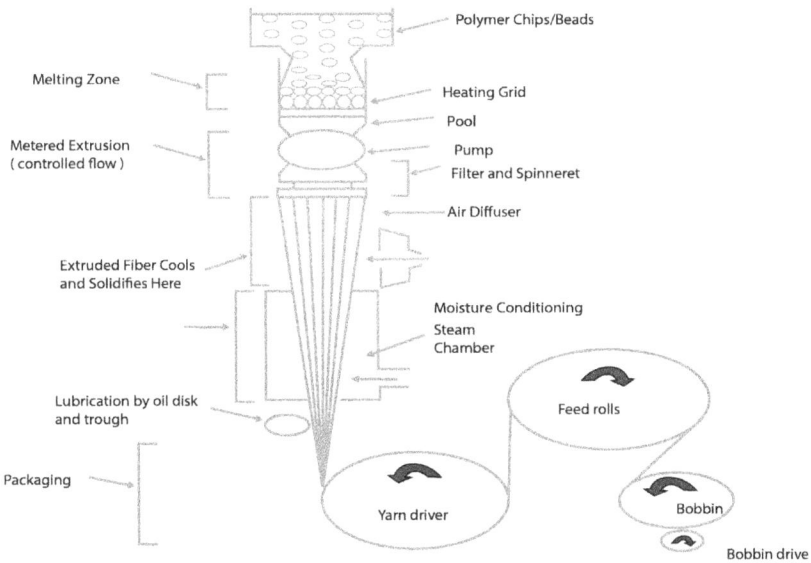

FIGURE 2.4   Melt fiber spinning process for specialty fibers.

then subjected to cooling in a fluid medium, which could be air, gas, or even water. This process facilitates the solidification of the filaments into fibers [25]. The process of melt spinning for extruding specialty fibers is shown in Figure 2.4.

2.3.2   SOLUTION SPINNING

Solution spinning is a technique employed when a polymer cannot be melted to form a stable melt. Instead, the polymer is dissolved in a solution to achieve a liquid state. There are two primary forms of solution spinning: dry spinning and wet spinning.

2.3.2.1   Dry Spinning

Dry spinning is a process that involves dissolving a polymer in a solvent that can easily evaporate. Once the polymer is dissolved, the resulting solution is then passed through a spinneret that is located in a drying tower that is sealed. The solution is then subjected to a drying process within the tower, during which the solvent evaporates, leaving behind a solidified polymer fiber (Figure 2.5) [26].

2.3.2.2   Wet Spinning

Wet spinning involves dissolving the polymer in a non-volatile solvent and then passing it through a spinneret into a coagulating bath, resulting in the precipitation of the fiber (Figure 2.6). However, high-performance fibers in this category can also be created using the simpler melt spinning technique with specialized drawing and annealing/carbonization processes. Another method involves two-stage drawing of a conventional melt-spun yarn, followed by an annealing step after each stage.

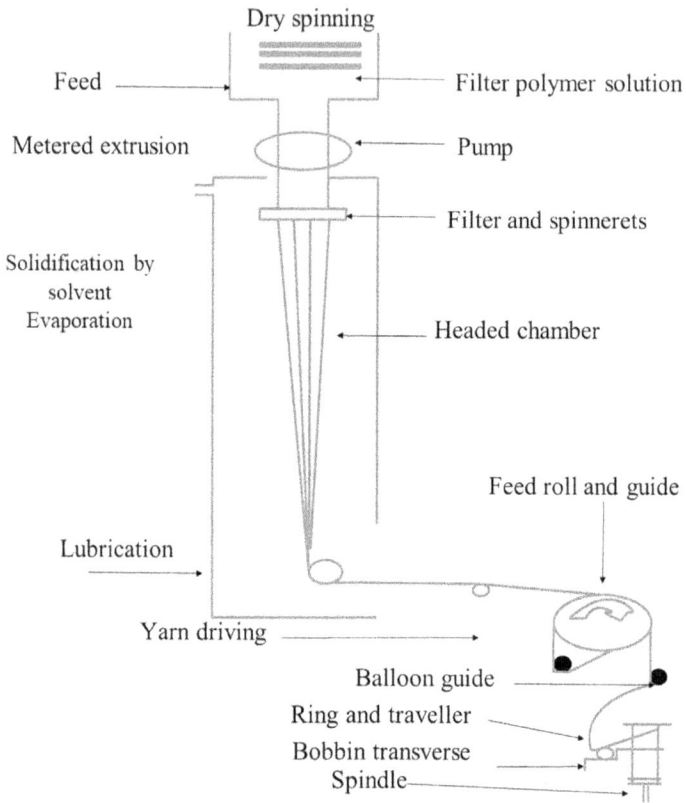

FIGURE 2.5   Dry wet spinning process for specialty fibers.

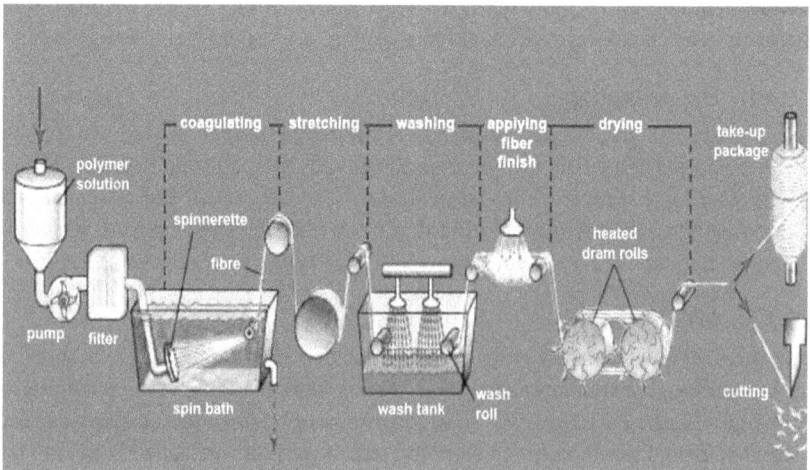

FIGURE 2.6   Wet spinning process of specialty fibers.

2.3.3 DRY JET–WET SPINNING

Dry jet–wet spinning is a modified version of the wet spinning method, in which the spinneret is placed just above the surface of the coagulation bath. During the process, the fiber is first extruded into an environment consisting of gas or air and then subsequently drawn into the coagulation bath. The bath is designed in such a way as to facilitate the vertical movement and coagulation of the fiber before it reaches the guide rollers. This technique enables the production of high-quality fibers with desirable physical properties [27]. Figure 2.7 illustrates the full process of Kevlar fiber manufacturing process via dry jet–wet spinning.

2.3.4 GEL SPINNING

Gel spinning technology is a widely used method for producing high-strength and high-mechanical-property fibers. This method, also known as semi-melt spinning, involves preparing gel-state fibers by extruding a polymer solution or plasticized gel from spinnerets, which is then cooled in solvent or water before being stretched into gel fiber using ultra-high extension. The most well-known commercially produced high-performance polyethylene fibers are Dyneema by DSM High Performance Fibers in the Netherlands and Spectra by Honeywell (formerly Allied Signal or Allied Fibers) in the USA, along with a Toyobo/DSM joint venture in Japan. The idea of creating a super-strong polyethylene fiber was first introduced in the 1930s by Carothers, but it was not until almost 50 years later that HDPE fibers were produced. The key to producing strong fibers is to stretch, orient, and crystallize the molecular chains of the polymer in the direction of the fiber. In regular polyethylene, the molecules are not oriented and are easily torn apart, but by using ultra-high-molecular-weight polyethylene, which has long

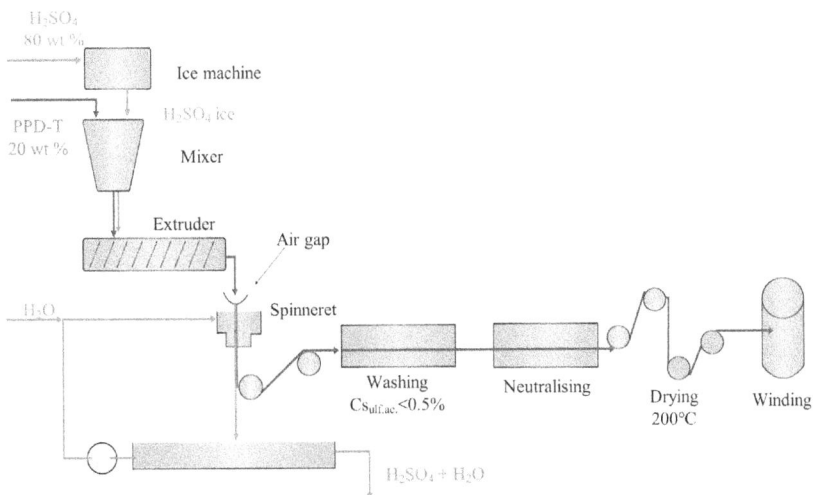

FIGURE 2.7 Dry jet–wet process for specialty fibers (Kevlar).

molecular chains that can interact with each other, it is possible to create fibers with exceptional strength [28, 29].

Typically, the extension and orientation of fibers are achieved through drawing, but this method is not viable for spinning melt-processed UHMW-PE due to its extremely high melt viscosity and the significant degree of molecular chain entanglement. However, these issues can be overcome through the gel spinning process, where the UHMW-PE molecules are dissolved in a solvent and spun through a spinneret. This process allows the molecules to become disentangled in the solution and maintain this state after being spun and cooled into filaments. The resulting gel-spun material has a low degree of entanglement and can be drawn to a high extent, known as super drawing. During the super drawing process, the macromolecules become highly oriented, resulting in a fiber with exceptional tenacity and modulus, as illustrated in Figure 2.8.

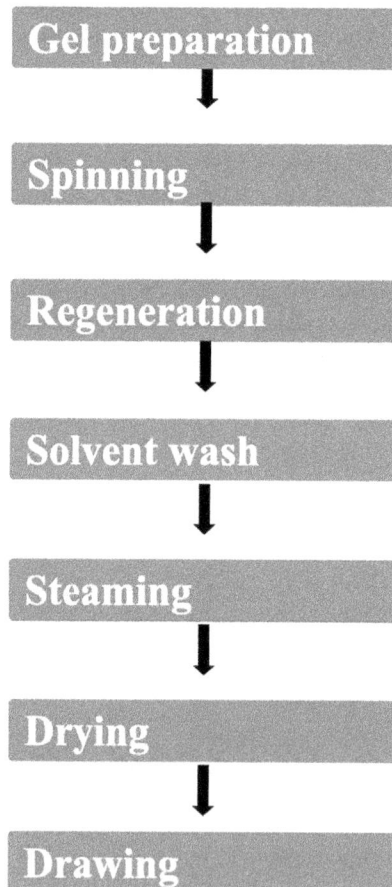

FIGURE 2.8 Flow diagram of gel spinning process and technology.

2.4 BRIEF SYNOPSIS OF FRONTIER FIBERS

2.4.1 CARBON FIBER

Carbon fiber is a popular choice for producing lightweight composite materials due to its high strength, elastic modulus, and low density. Polyacrylonitrile fibers are an excellent precursor for carbon fibers, and their transformation into carbon fibers involves two stages, stabilization and carbonization. Stabilization requires heating the PAN fibers in an oxygen-rich environment to further orient and crosslink the molecules to withstand high-temperature pyrolysis without decomposition. Carbonization involves heat treatment in an inert atmosphere to remove non-carbon elements in the form of gases such as CO_2, H_2O, NH_3, HCN, CO, and N_2. The quality of the final carbon fibers largely depends on the precursor fibers and various process parameters used during stabilization and carbonization. Zoltek Corp., located in St. Louis, is a leading manufacturer of carbon fibers and has developed Pyron, a fiber suitable for flame-retardant fabrics. Pyron is an oxidized polyacrylonitrile fiber that combines the high-temperature and flame resistance of carbon fibers with the manageable processability required for textile operations. Pyron is inherently fire resistant and thus ideal for use in protective fabrics [12, 30].

2.4.2 ULTRA-HIGH-MOLECULAR-WEIGHT POLYETHYLENE

Honeywell's Spectra S3000 fiber, produced from ultra-high-molecular-weight polyethylene, using a patented gel-spinning process is being used for ballistic applications. The materials provide up to 20% greater ballistic performance than the company's first Spectra Shield line, according to the company. Fibers for structural or industrial applications require high strength, high stiffness, and other properties, depending upon the ultimate end use [31]. Magellan Systems International, Arnhem (NL), has developed a new type of synthetic fiber, M5, which possesses greater tenacity than para-aramid or carbon fibers and a higher modulus than PBO fibers.

2.4.3 POLYBENZIMIDAZOLE FIBERS

Polybenzimidazole (short for poly[2,2'-(m-phenylen)-5,5'-bisbenzimidazole]) fibers possess exceptional textile characteristics and can be easily processed using conventional textile equipment. Their properties make them highly suitable for blending with other fibers, resulting in the creation of high-quality products that exhibit outstanding flame resistance, as well as a soft and comfortable texture similar to cotton. Moreover, PBI fibers can enhance the processability of their partnering fiber in numerous instances [32].

2.4.4 POLY(ETHER IMIDE)

Poly(ether imide) or PEI is an engineering plastic that shares similarities with PEEK in terms of its applications as a fiber material in industries that require high levels of chemical and temperature resistance. However, while its temperature resistance may be slightly inferior to that of PEEK, PEI is a more cost-effective option. Leading companies such as Teijin and Acordis (formerly Akzo) have successfully developed and commercialized PEI-based fibers, with both products believed to have similar chemical structures [33].

2.4.5 Fluorinated Fibers: PTFE, PVF, PVDF, and FEP (ARH)

Fluoropolymeric fibers are known for their high cost, but their exceptional chemical inertness makes them ideal for use in filtration applications that require resistance to harsh chemicals and sometimes high temperatures [34]. Poly(tetrafluoroethylene) is the most well-known example of such fibers, but other types include poly (vinyl fluoride), poly(vinylidene fluoride), and various fluorinated ethylene polymers.

2.4.6 Silicon Carbide-Based Fibers

The demand for strong reinforcements in ceramic matrix composites (CMCs) and metal matrix composites (MMCs) for high-temperature applications above 1000 °C has driven significant advancements in the production of small-diameter ceramic fibers [35]. Fine-diameter silicon carbide (SiC) fibers are produced from precursor fibers that are spun from organosilicon polymers, such as polycarbosilane (PCS) or its derivatives. These polymers are composed of six-atom rings of Si and C, which mimic the blend structure of SiC. Methyl groups and hydrogen atoms are attached to these rings and remain in the fibers even after the fibers are converted to ceramic through pyrolysis above 1200 °C. However, before this conversion process can occur, crosslinking of the precursor fiber is necessary to prevent the fiber from softening or melting during pyrolysis. The selection of precursor polymers for SiC fibers and the crosslinking processes used have a significant impact on the final composition and microstructure of the ceramic fibers.

2.4.7 Glass Fibers

Fiberglass or glass fiber is a widely used reinforcement in the polymer industry. It is highly versatile and can be made into sheets or randomly woven into fabrics. Fiberglass is composed of inorganic glasses based on silica, which forms three-dimensional networks through polymerization. Fiberglass is classified based on the constituent raw materials and their composition. The classifications include D-glass, E-glass, ECR-glass, R-glass, S-glass, and S-2 glass. E-glass is the most commonly used fiber in the fiber-reinforced polymer composite industry due to its strength and electrical resistance. S-glass and S-2 glass are high-strength glasses used where extreme temperature and corrosive resistance is required. S-2 glass is a brand name originally created by Owens-Corning but is now a registered trademark of AGY Holdings Corp [36].

2.4.8 NYLON

Nylon is a versatile synthetic fiber that offers several desirable characteristics, such as lightweight construction, excellent strength, and a soft texture that is highly durable. Additionally, it exhibits fast-drying capabilities, making it an ideal material for use in conjunction with polyurethane coatings. Compared to polyester, nylon has a much higher moisture regain, allowing it to wick moisture away more effectively. This makes it an excellent choice for use in tightly woven outerwear, where its low air permeability can trap heat. However, it is also well suited for use in more breathable knitted fabrics, where it can provide exceptional performance.

2.4.9 POLYESTER

Polyester is an ideal choice for base fabrics in active wear due to its exceptional dimensional stability and resistance to dirt, decay, mold, and most organic solvents, as well as its outstanding thermal stability. Additionally, it is a cost-effective option that has low moisture absorption and requires minimal care. While polyester is generally hydrophobic and doesn't absorb moisture, polyester used in base-layer clothing is typically treated with chemicals to enhance its ability to wick moisture. In contrast, viscose rayon is not typically used next to the skin in sportswear, as it retains water (with a moisture regain of 13%). However, it can be incorporated as the outer layer of knitted hydrophilic twin-layer sportswear, as it can absorb 2–3 times more moisture than cotton. By incorporating hydrophilic finishes, the wicking behavior can be further improved.

2.4.10 POLYPROPYLENE

Polypropylene cannot wick liquid moisture. However, moisture vapor can still be forced through polypropylene fabric by body heat. Polypropylene upon melting has the advantage of providing superior thermal insulation properties even in the wet conditions. Polypropylene is claimed to be a proven performer in moisture management due to its hydrophobic nature and has very good thermal characteristics, keeping the wearer warm in cold weather and cold in warm weather.

2.5 SPECIAL COMMERCIAL FIBERS

Table 2.4 summarizes the commercially available specialty fibers. A few examples are illustrated in this section.

2.5.1 HYGRA 20

Unitika Limited recently introduced a new filament yarn called Hygra, which consists of a unique combination of water-absorbing polymer and nylon. The sheath

TABLE 2.4

Functional and High-Performance Specialty Fibers and Their Global Status

Fiber/Brand	Companies	Applications
Meta-aramids	Nomex DuPont (USA), Teijin Twaron (Japan), SRO Group (China), Yantai Spandex (China), Kermel (France)	Bulletproof jackets, body armor, helmets, and NBC outer fabrics
Para-aramids	Du Pont (Kevlar), Teijin Twaron (Japan), Yantai Spandex (China)	Ballistic fabrics and flame-resistant clothing
Modacrylic fibers	Solutia Inc (USA), Kaneka Corporation (Japan), Yalova Eliat (Turkey), Montefibre (Italy) & Mosanto (USA)	Fire-resistant blankets, curtains, carpets, sofa covers
Super-absorbent fibers (acrylics)	Technical Absorbent Ltd. (UK), Technical Absorbent Ltd. (UK), Camelot Technologies (Canada)	Feminine hygiene products and inner layer of DPT–NBC and HAPG
Ultra-high-molecular-weight polyethylene	Taniyama Chemical Industry (Japan), Royal DSM (Netherlands), and Honeywell (USA)	Bulletproof vests, ballistic body armor, and base fabric for aerostats
Carbon fibers	Toray Industries (Japan), Toho Tenax (Japan), Mitsubishi Rayon (Japan), Zoltek (USA), Hexcelcorp (USA) & SGL Carbon AG (Germany).	CBRN adsorbent materials
Fire-retardant viscose	Shandong Helon Textile Sci. & Tech. Co. Ltd (China) & Lenzing AG (Austria)	CBRN outer and inner layers
Flame-retardant polyester	DuPont, Shanghai Jingmao Industrial Co. Ltd (China), Aquafil Spain (Italy) & Reliance Trevira	Firefighter suits and CBRN outer layer
High-tenacity/super high–tenacity nylon	Junma (China) & Kordsa (Turkey)	Inftabel applications
High-tenacity/super high–tenacity polyester	Fibres (USA), Teijin Twaron (Japan), Toray Industries (Japan), Hyosung Corp & Reliance	Heavy textiles and CBRN fabrics
High-tenacity/super high–tenacity polypropylene	DuPont (USA) & Drake (Fibres) Ltd (UK)	UV protective textiles/parachutes
Polytetrafluoroethylene	DuPont (USA), Newton Filaments, Inc (USA), Albany International Inc. (USA) & Toyobo (Japan)	Heavy textiles and barrier layer for CBRN

Fiber/Brand	Companies	Applications
PBO	Toyobo Co. Ltd. (Japan)	High tensile strength and high flame resistance properties
Anti-microbial/anti-fungal/ anti-bacterial fibers	Trevira, Montefibre, Brilen, Sterling, Kaneba & Zimmer AG	CBRN, heavy textiles, inner layer CWPC, combat desert wear, protective clothing for doctors, adverse climatic conditions
Phenolic fibers	Phenco (USA), The Vermont Organic Fiber Company (USA),	Automotive and electrical components,
Conductive fibers	Shakespeare LLC	electronics manufacturing garments, clean room garments, military garments
Multifunctional fibers	Reliance Industries Ltd. India	CBRN—Outer layer

core design of Hygra enables it to have excellent water absorption properties due to the special network structure of the water-absorbing polymer used, which can absorb up to 35 times its weight in water [37]. Furthermore, Hygra has the ability to quickly release absorbed moisture, which sets it apart from conventional water-absorbing polymers. Another notable feature of Hygra is its exceptional antistatic properties, which remain effective even in low-wet conditions. These attributes make Hygra an ideal choice for various apparel applications such as sportswear, including athletic wear, skiwear, and golf wear.

2.5.2 Killat N23

Killat N, a product of Kanebo Ltd, is a type of nylon filament that has a unique feature of being hollow, with approximately 33% of its cross-section consisting of empty space. This characteristic of the filament allows it to have excellent water absorbency and thermal insulation properties. The manufacturing process of Killat N is quite intriguing. It involves spinning a bicomponent filament yarn, which comprises a soluble polyester copolymer as the core component and nylon as the outer component or skin.

2.5.3 Lycra25

Lycra is a synthetic fiber made up of long-chain polymers that consist of at least 85% segmented polyurethane. It is widely used in various end-use applications such as swimwear, active sportswear, and floor gymnastics wear due to its superior comfort and fit. When added to a fabric, Lycra provides stretch and recovery, making it ideal for activities such as gymnastics and swimming where the body is constantly moving and stretching. In comparison to other types of Lycra, such as Lycra T-9026, this variant requires less effort to achieve the same level of extensibility.

2.5.4 DACRON

Polyester 4-Channel is a versatile and advanced type of fiber designed to effectively manage moisture by accelerating the evaporation of sweat. This fabric excels in its ability to wick moisture away from the skin, promote rapid drying, and facilitate efficient moisture absorption and transportation.

2.5.5 REGENERATED TENCEL FIBER

Lyocell, also known by the generic name Tencel, is a type of natural and human-made fiber that is produced through an eco-friendly process using wood pulp. It has gained popularity in clothing due to its unique moisture management properties, particularly in sports performance. Tencel's exceptional ability to absorb moisture makes it perfect for the skin, promoting a high level of comfort and overall well-being [38].

2.5.6 BAMBOO

Bamboo fabric is created using pure bamboo fiber yarns, offering a range of superior qualities such as excellent moisture permeability, softness, and excellent draping. Additionally, it is effortless to dye, providing a range of beautiful colors. Bamboo fabric is a new and highly promising eco-friendly textile. It possesses exceptional antibacterial function, making it an ideal choice for producing items such as undergarments, form-fitting t-shirts, and socks. The fabric's inherent UV-resistant nature also makes it an excellent choice for creating summer clothing [39, 40].

2.5.7 SOYBEAN

The protein found in soybean fiber can create a high-quality, soft texture that offers both moisture absorption and permeability, making it ideal for use in knitted fabrics and undergarments. Furthermore, when treated with an antibacterial agent, it can also provide health-related benefits. This material shows promise for use in premium-quality knitted fabrics and undergarments.

2.6 PROPERTIES

In this section, the properties of various technical-grade fibers are presented in Table 2.5.

Fibers' properties can be divided into two groups; those with elastic moduli lower than the cement matrix, such as cellulose, nylon, and polypropylene, and those with higher elastic moduli, such as asbestos, glass, steel, and carbon. Another classification based on fiber characteristics can be made according to the origin of the fiber material, such as metallic, polymeric, or natural. High-performance fibers possess remarkable strength and thermal stability, making them suitable for a wide range of cutting-edge applications. These fibers are extensively utilized in

TABLE 2.5
Properties of Technical-Grade Fibers

Type of Fiber	Tensile Strength (MPa)	Young's Modulus (GPa)	Ultimate Elongation (%)	Specific Gravity
Acrylic	210–420	2.1	25–45	1.1
Asbestos	560–980	84–140	0.6	3.2
Carbon	1800–2600	230–380	0.5	1.9
Glass	1050–3850	70	1.5–3.5	2.5
Nylon	770–840	4.2	16–20	1.1
Polyester	735–875	8.4	11–13	1.4
Polyethylene	700	0.14–0.42	10	0.9
Polypropylene	560–770	3.5	25	0.9
Rayon	420–630	7	10–25	1.5
Rock wool	490–770	70–119	0.6	2.7
Steel	280–2800	203	0.5–3.5	7.8

various sectors, including aerospace textiles, biomedical devices, civil engineering, construction, protective clothing, geotextiles, and electronics [12]. One of the crucial characteristics of high-performance fibers is their resistance to high temperatures and flames, which plays a crucial role in determining their operational conditions. High-performance fibers are designed to meet specific requirements for exceptional strength, stiffness, heat resistance, or chemical resistance. These fibers have higher tenacity and modulus compared to typical fibers and are available in a wide range of properties, as shown in Table 2.4. Although high-performance fibers are generally niche products in the fiber market, some are produced in large quantities. Glass is the oldest high-performance fiber and is used in insulation, fire-resistant fabrics, and fiberglass composites. Carbon fiber is crucial for military and aerospace applications due to its strength and stiffness, and variations can have different electrical conductivity, thermal, and chemical properties. Organic fibers, such as aramids, have also become essential and exhibit high tensile strength and thermal resistance. Para-aramids have the highest impact resistance, making them popular for body armor, and can be blended with other fibers for less demanding applications. The manufacturing process for high-performance fibers includes extruding an organic precursor material into filaments and then carbonizing them to convert them into carbon, with different precursors besides carbonization processes are used for desired properties [12, 13]. Important fiber properties of different technical-grade fibers are listed in Table 2.4.

2.7 APPLICATION OF EMERGING FIBERS

High-modulus, high-tenacity (HM-HT) fibers can be categorized into three groups based on their composition: polymer fibers (e.g. aramids and polyethylene), carbon fibers (e.g. Kevlar), and inorganic fibers made of glass and ceramic. These fibers are becoming increasingly popular for a variety of applications, including geotextiles, geomembranes, construction, civil engineering projects, and composite materials. Their exceptional properties make them an excellent choice for engineers and materials scientists looking for durable and reliable materials. Specialist fibers within composite materials can also benefit from the use of high-performance fibers, as they are capable of fulfilling challenging roles in each area of technical textile segments. Utilization of fibers in various fields are found in diverse areas, as shown in Table 2.6.

TABLE 2.6
High-Performance/Functional Fibers/Specialty Textile Fiber Product Applications in Each Segment of Technical Textiles

Description	Broad Market Areas	Products/Applications
Agrotech	Agriculture, aquaculture, horticulture, and forestry	Woven/nonwoven crop protection covers, capillary matting, land netting, fishing ropes, fishing nets, fishing line, baler twine
Buildtech	Building and construction	Tarpaulins, textile structures, awnings, roofing felts, sewer linings, woven roofing, roof scrims, house wrap, hoardings, scaffold nets, concrete reinforcement (fiber, scrims), composites
Clothtech	Technical components of footwear and clothing	Woven/nonwoven interlinings, waddings, laces, shoe components, sewing threads, hook and loop fasteners, zips/other fasteners, labels
Geotech	Geotextiles for civil engineering	Ground stabilization, soil reinforcement, pit linings, erosion control
Hometech	Technical components of furniture, household textiles, and floor coverings	Woven/nonwoven wipes, vacuum filters, HVAC filters, pillow ticking, mattress ticking, mattress components, spring wraps, spring insulators, platform cloths, dust cloths, fiberfill, furniture components, webbings, curtain tapes, woven or nonwoven primary or secondary carpet backing
Indutech	Filtration, conveying, cleaning, and other industrial uses	Conveyor belting; hoses; drive belting; brushes; battery separators; other electrical goods; abrasives; PCBs; electrical composites; woven filters; cable components; nonwoven air, dust, liquid, or other filters; cigarette filters; paper-making felts; woven/nonwoven wipes; gaskets; fiberfill; anti-corrosion composites; lifting webs; ropes; silos; oilbooms; other coating substrates
Medtech	Hygiene and medical	Woven/nonwoven gowns and drapes, woven/knit/nonwoven wound care, sterile packaging, medical mattresses, cover stock, cotton wool, wipes
Mobiltech	Automobiles, ships, aircraft, and railways	Car/truck tire cords, drive belts, hoses, cabin filters, seat belts, air bags, tufted or needled carpet, carpet backing, woven/knit or nonwoven trim, upholstery, insulation, truck covers, ropes, transport, marine composites

Description	Broad Market Areas	Products/Applications
Oekotech	Environmental protection	House wraps, erosion control, pit linings, woven filters, nonwoven dust filters, automotive insulation
Packtech	Packaging	Flexible intermediate bulk containers (FJBCs), laundry bags, sacks, twine, teabags, food soaker pads, other nonwoven packaging, netting, fiber strapping, other woven packaging
Protech	Personal and property protection	Fire retardance, various defense protective textiles nuclear biological chemical (NBC) resistant, cut/slash resistant, bulletproof, heat/cold/chemical/radiation protective, hi-visibility harnesses, clothing, face masks, gloves
Sporttech	Sports and leisure composites	Boat covers, book cloth, shopping bags, sports bags/luggage, leather substrates, sailcloth, artificial turf, ropes, nets, balls, flags, air-sport fabrics, tents, sleeping bag fill, fabrics, webbing, equipment

2.8   SPINNABLE GRADES OF POLYMERIC MATERIALS

Polymer, which means "many units", refers to a macromolecule formed by linking smaller molecules or monomers. In the technical textile industry, spinnable-grade polymers are essential precursors used in the fiber spinning process. These polymers are either melted or dissolved and then extruded through a spinneret before being drawn to obtain technical-grade fibers. The demand for such polymeric fibers is increasing due to the flexibility of processing, which allows the material to be tailored to specific mechanical, electrical, optical, and chemical properties required for a particular application [2]. Figure 2.9 provides an approach to classifying fiber-forming polymers based on the chemical structure of the fibers, highlighting their significance as a special class of polymeric materials.

1. Organic polymers are a type of polymer that consists of a carbon backbone chain, either aliphatic or aromatic. Examples of spun fibers from these polymers include aramid, polyester, polyamide, acrylic, modacrylic, chlorofiber, vinylal, fluorofiber, polyethylene, and polypropylene. Organic polymers can be classified into two types: flexible-coil fibers that have an alkane backbone, such as HDPE, and rigid-rod fibers that have aromatic moieties in the main chain. These aromatic fibers can create highly ordered states in the melt or solution, such as lyotropic liquid crystalline polymers (Kevlar, Nomex, Technora, Zylon) and thermotropic liquid crystalline polymers (Vectran).

2. In organic polymers, the backbone is constituted by elements other than carbon, such as boron, glass, carbon, metal, and silicon carbide (ceramic).

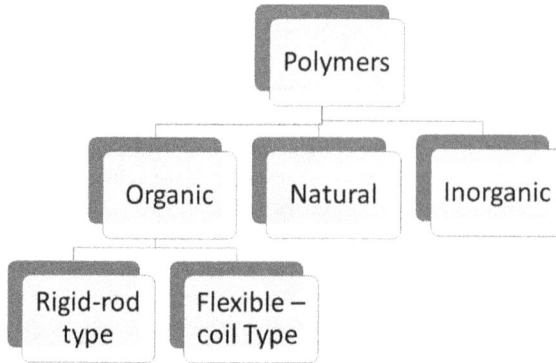

FIGURE 2.9 Classification of spinnable grades of fiber polymer.

3. Natural polymers are fibers spun from naturally derived products, such as acetate (CA), triacetate (CTA), alginate (ALG), viscose (CV), modal (C11D), lyocell (CLY), and chitosan.

Table 2.7 presents a summary of different types of polymeric materials (precursors) that can be spun into fibers, including their respective trade names.

TABLE 2.7

Polymeric Precursors and Their Respective Trade Names

Polymer Precursor Name	Common Trade Name
Regenerated cellulose	Rayon
Cellulose triacetate	acetate, Arnel
Polycaprolactam (textile fiber)	Nylon 6 (textile)
Polyhexamethylene adipamide (textile fiber)	Nylon 66
Poly-p-phenylene tereph-thalamide	Kevlar, Twaron, Technora
Poly-m-phenylene isoph-thalamide	Nomex, Conex
Polyethylene terephthalate	Dacron, Terylene, Trevira
Acrylic (>85% acrylonitrile)	Acrilan, Creslan, Courtelle
Modacrylic (35–85% acrylonitrile)	Verel
Polyurethane	Lycra Stretchable fibers
Ultra-high-molecular-weight Polyethylene (Mn~ 5–10 millions)	Dyneema fiber, Spectra
Petroleum mesophase pitch–based carbon fibers	Union Carbide Corporation, USA
Coal tar mesophase pitch–based carbon fibers	Osaka Gas Co., Ltd. DONACARBO-F and—S Nippon Graphite Fiber Corp.
Acrylic-based carbon fibers	Toray Industries, Inc. Toray Industries, Inc., Torayca, PANEX and PYRON Mitsubishi's products (trade name PYROFI Solvay, Cytec Thornel, and Thermal Graph)

Polymer Precursor Name	Common Trade Name
Rayon	Teijin Ltd., Tokyo
Diaminodihydroxy benzene	Zylon (PBO)
Polycarbosilane	Silicon carbide (SiC) ceramic fibers
E-glass fibersalumino-borosilicate glass	Vitro-fibras
Basalt fibers	Kamenny Vek (Dubna, Russia), Sudaglass Fiber Technology Inc. (Houston, TX, US), and Technobasalt-Invest LLC (Kyiv, Ukraine)
Thermotropic liquid crystalline polymers (melt) polyester	Vectran

2.9 SPECIAL TECHNIQUES—BICOMPONENT FIBERS

Bicomponent fibers, also known as "conjugated fibers", are a type of composite fiber formed by two different polymers that are fed into separate spinning channels as melts or solutions and extruded through the same nozzle hole. These fibers can have various cross-sections depending on the shape of the spinneret used. Bicomponent fibers are prepared using methods such as spun bonding, electrospinning, melt-blowing, and gel spinning, which combine the advantages of the two raw materials to obtain fibers with diverse properties. The emergence of bicomponent fibers with excellent properties has had a significant impact on the textile industry. These fibers have enriched the range of available fiber types and improved the quality of prepared textiles. As a result, they have attracted the attention of scientific researchers. Bicomponent fibers offer benefits such as microdenier fiber production, utilization of special polymers or additives at reduced cost, uniform distribution of adhesive, maintaining fiber integrity within the structure, and providing functional integrity. Additionally, they are environmentally friendly, as they do not generate effluents, and they are recyclable.

The production technology of bicomponent fibers primarily focuses on filament, staple fibers, and heat-bonded nonwoven fabrics. Their applications extend to various fields, including medicine, energy, and clothing. The importance of bicomponent fibers continues to grow due to their multiple uses. Bicomponent fiber spinning involves a wide variety of diameters, often referred to as super-microfibers when they are 0.3 dtex (<1denier) or less in diameter. These fibers can generally be classified into five categories based on their cross-sectional shape: side-by-side, core-shell, eccentric, island-in-the sea, and segmented-pie types [41]. Figure 2.10 illustrates a straightforward depiction of different cross-sectional bicomponent fibers.

ES FIBERVISIONS offers a range of commercially available products with unique features. One such product is the PTC bicomponent fiber, which combines polyester and polypropylene. This fiber exhibits exceptional qualities, including excellent bulkiness, resilience, ultrasonic bondability, and processability under

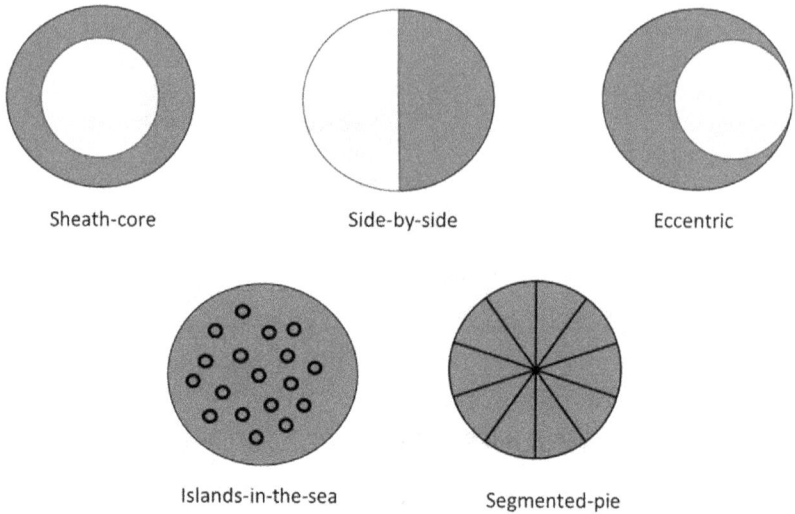

Sheath-core   Side-by-side   Eccentric

Islands-in-the-sea   Segmented-pie

FIGURE 2.10 Various cross-sectional product configurations of bicomponent fibers are displayed.

a wide range of conditions. It finds successful application in various industries, such as hygiene acquisition-distribution layers, filters, household goods, medical products, and through-air bonded products (oven-bonded), as well as carded thermal bonded products. Another notable product is the Soft Bicomponent fiber, also from ES FIBERVISIONS. It is primarily composed of polypropylene and polyethylene as its core and sheath materials, respectively. This bicomponent fiber is specifically designed to possess inherent softness, strong hydrophilicity, and wettability. The melting point difference between the polyethylene sheath (130 °C) and the polypropylene core (160 °C) enables effective bonding of the sheath while preserving the integrity of the core. Finally, ES FIBERVISIONS offers the PTC bicomponent fiber with outstanding characteristics for various applications, and the Soft Bicomponent fiber, which combines softness, hydrophilicity, and wettability through its unique polypropylene/polyethylene composition. In summary, bicomponent fibers are composite fibers formed by two different polymers and offer a wide range of properties and applications. They have diverse cross-sectional shapes and are produced using various spinning methods. These fibers have contributed to advancements in the textile industry and gaining importance due to their numerous benefits and applications.

2.10 CASE STUDY

Gel spinning of polyethylene is a special process used to obtain high strength, high modulus polyethylene. A polymer is dissolved in 2, 3, and 5 %w/w in decalin with slow heating and continuous stirring at 150 °C. The other process parameters are kept constant for making a comparative analysis. Apart from the conventional melt spinning process used for thermoplastic materials, gel spinning of PE is a solution spinning process where ~1–5 wt% of UHMWPE (Mw ~ 1–5 × 106) gel in decalin or paraffin oil is extruded at 130–150 °C from a spinneret through an air gap to a cooling bath (cold water) and further drawn in hot air oven in the 90–120 °C to produce ultra-high-molecular-weight polyethylene fibers. Prior to the spinning process, polyethylene is dissolved in decalin or other appropriate solvent. Polyethylene gel typically melts at ~150 °C. The fibers obtained after the cooling bath, generally called gel fibers, contain most of the solvents in an unoriented isotropic material. However, due to the presence of an inter- and intramolecular entangled mesh-like structure, it can still withstand a good amount of stress and effectively transmit forces during further drawing process [42]. Hence, it allows the fiber as spun to be drawn as many as 100 times or higher. By adopting this method successfully, the spinning of fibers with a tenacity of ~30–50 gpd and Young's modulus as high as ~ 1000–1500 gpd is possible.

2.11 CONCLUSION

The time has come to move from basic commodity textiles to high-value textile items, which is only possible with vigorous R&D as long-term policy. This chapter presents findings for eco-friendly and sustainable technlgies and the concept of new technical fibers. Engineering of fibers with specific characteristics requires a thorough understanding of polymers, fibers, and processing science. Above stated points are covered with some interesting examples of various materials for manufacturing technical grade fabric: (a) Modified materials like bi-component fibers, high-performance functional fibers like conductive, hybrid fibers, superabsorbent fibers, high strength and high modulus fibers, and UV protective fibers. (b) Multifunctional polymers spun into fiber with several functional properties-antistatic, flame and chemical resistant synthesized in one fiber. (c) Textile performance chemical and multifunctional layered film coated fabric structure is other segment of technical textiles for making innovative durable impermeable fabric for chemical storage and various impermeable applications. Exploring these high-performance fibers will enhance the shelf life of technical textile fabrics and thereby ensure the stability of technical textiles.

REFERENCES

1. Schmidt-Rohr K, Clauss J, Spiess HW. 1992. Correlation of structure, mobility, and morphological information in heterogeneous polymer materials by two-dimensional wideline-separation NMR spectroscopy. Macromolecules 25(12):3273–3277. https://doi.org/10.1021/ma00038a037
2. Sinha MK, Das BR, Kumar K, Kishore B, Prasad NE. 2017. Development of ultraviolet (UV) radiation protective fabric using combined electrospinning and electrospraying technique. Journal of The Institution of Engineers (India): Series E 98(1):17–24. https://doi.org/10.1007/s40034-017-0094-z
3. Paul R. 2019. High performance technical textiles: An overview. In High Performance Technical Textiles. John Wiley & Sons, West Sussex, UK, 1–10.

4. Sinha MK, Das BR, Prasad N, Kishore B, Kumar K. 2018. Exploration of nanofibrous coated webs for chemical and biological protection. Zaštita Materijala 59(2):189–198. https://doi.org/10.5937/ZasMat1802189K

5. Datta R, Henry M. 2006. Lactic acid: Recent advances in products, processes and technologies—A review. Journal of Chemical Technology & Biotechnology: International Research in Process, Environmental & Clean Technology 81(7):1119–1129. https://doi.org/10.1002/jctb.1486

6. Pandit P, Pandey R, Singha K, Shrivastava S, Gupta V, Jose S. 2020. Pineapple leaf fibre: Cultivation and production. In M. Jawaid et al. (eds) Pineapple Leaf Fibers, Green Energy and Technology. Springer Nature, Singapore, 1–20. https://doi.org/10.1007/978-981-15-1416-6_1

7. Pandey R, Pandit P, Pandey S, Mishra S. 2020. Solutions for sustainable fashion and textile industry. In Pintu Pandit et al. (eds) Recycling from Waste in Fashion and Textiles: A Sustainable & Circular Economic Approach. Scrivener Publishing, 33–72. https://doi.org/10.1002/9781119620532.ch3

8. Pandey R, Dubey A, Sinha MK. 2023. Lotus fibre drawing and characterization. In R Nayak (ed.) Sustainable Fibres for Fashion and Textile Manufacturing. Woodhead Publishing, 95–108. https://doi.org/10.1016/B978-0-12-824052-6.00001-9

9. Pandey R, Sinha MK, Dubey A. 2023. Macrophyte and wetland plant fibres. In R Nayak (ed.) Sustainable Fibres for Fashion and Textile Manufacturing. Woodhead Publishing, 109–127. https://doi.org/10.1016/B978-0-12-824052-6.00006-8

10. Rao Y, Farris RJ. 2000. A modeling and experimental study of the influence of twist on the mechanical properties of high-performance fiber yarns. Journal of Applied Polymer Science 77(9):1938–1949. https://doi.org/10.1002/1097-4628(20000829)77:9%3C1938::AID-APP9%3E3.0.CO;2-D

11. Hearle JW. 2001. High-Performance Fibres. Woodhead Publishing, Cambridge, England.

12. Dasaradhan B, Das BR, Sinha MK, Kumar K, Kishore B, Prasad NE. 2018. A brief review of technology and materials for aerostat application. Asian Journal of Textile 8(1):1–12. https://doi.org/10.3923/ajt.2018.1.12

13. Chavan S, Kanu NJ, Shendokar S, Narkhede B, Sinha MK, Gupta E, Singh GK, Vates UK. 2023. An insight into nylon 6, 6 nanofibers interleaved E-glass fiber reinforced epoxy composites. Journal of the Institution of Engineers (India): Series C 104(1):15–44. https://doi.org/10.1007/s40032-022-00882-0

14. Pandey R, Prasad GK, Dubey A, Arputhraj A, Raja ASM, Sinha MK, Jose S. 2022. Tellicherry bark microfiber: Characterization and processing. Journal of Natural Fibers 19(16):13288–13299. http://dx.doi.org/10.1080/15440478.2022.2089432

15. Caicedo C, Melo-López L, Cabello-Alvarado C, Cruz-Delgado VJ, Ávila-Orta CA. 2019. Biodegradable polymer nanocomposites applied to technical textiles: A review. Dyna 86(211):288–299. https://doi.org/10.15446/dyna.v86n211.80230

16. Badanayak P, Jose S, Bose G. 2023. Banana pseudostem fiber: A critical review on fiber extraction, characterization, and surface modification. Journal of Natural Fibers 20(1):2168821. https://doi.org/10.1080/15440478.2023.2168821

17. Pandey R. 2016. Fiber extraction from dual-purpose flax. Journal of Natural Fibers 13(5):565–577. https://doi.org/10.1080/15440478.2015.1083926

18. Pandey R, Jose S, Basu G, Sinha MK. 2021. Novel methods of degumming and bleaching of Indian flax variety tiara. Journal of Natural Fibers 18(8):1140–1150. https://doi.org/10.1080/15440478.2019.1687067

19. Pandey R, Tiwari N, Dubey A, Jose S, Kambo N, Joshi S, Chauhan VK, Basu G. 2022. A comparative study of phenotypic variability and physico-mechanical properties of dual-purpose flax fiber varieties in India. Journal of Natural Fibers 19(17):15680–15689. https://doi.org/10.1080/15440478.2022.2133048

20. Pandey R, Sinha MK, Dubey A. 2020. Cellulosic fibers from lotus (Nelumbo nucifera) peduncles. Journal of Natural Fibers 17(2):298–309. https://doi.org/10.1080/1544047 8.2018.1492486

21. Pandey R, Mishra S, Dubey R. 2023. Luxurious sustainable fibers. In SS Muthu (ed.) Novel Sustainable Raw Material Alternatives for the Textiles and Fashion Industry. Springer Nature Switzerland, Cham, 57–79. https://doi.org/10.1007/978-3-031-37323-7_4

22. Pandey R, Jose S, Sinha MK. 2022. Fiber extraction and characterization from Typha domingensis. Journal of Natural Fibers 19(7):2648–2659. https://doi.org/10.1080/154 40478.2020.1821285

23. Brazinsky I, Williams AG, LaNieve HL. 1975. The dry spinning process: Comparison of theory with experiment. Polymer Engineering & Science 15(12):834–841.

24. Tavanaie MA. 2021. Engineered biodegradable melt-spun fibers. In Engineered Polymeric Fibrous Materials. Woodhead Publishing, 191–232. https://doi.org/10.1016/B978-0-12-824381-7.00014-7

25. Hufenus R, Yan Y, Dauner M, Kikutani T. 2020. Melt-spun fibers for textile applications. Materials 13(19):4298. https://doi.org/10.3390/ma13194298

26. Imura Y, Hogan RMC, Jaffe M. 2014. Dry spinning of synthetic polymer fibers. In Advances in Filament Yarn Spinning of Textiles and Polymers. Woodhead Publishing, 187–202. https://doi.org/10.1533/9780857099174.2.187

27. Gao Q, Jing M, Wang C, Chen M, Zhao S, Wang W, Qin J. 2019. Correlation between fibril structures and mechanical properties of polyacrylonitrile fibers during the dry-jet wet spinning process. Journal of Applied Polymer Science 136(14):47336. https://doi.org/10.1002/app.47336

28. Yufeng Z, Changfa X, Guangxia J, Shulin A. 1999. Study on gel-spinning process of ultra-high molecular weight polyethylene. Journal of Applied Polymer Science 74(3):670–675. https://doi.org/10.1002/(SICI)1097-4628(19991017)74:3%3C670::AID-APP21%3E3.0.CO;2-8

29. Xia L, Xi P, Cheng, B. 2015. A comparative study of UHMWPE fibers prepared by flash-spinning and gel-spinning. Materials Letters 147:79–81. https://doi.org/10.1016/j.matlet.2015.02.046

30. Ma Z, Song H, Wang H, Xu P. 2017. Improving the performance of microbial fuel cells by reducing the inherent resistivity of carbon fiber brush anodes. Journal of Power Sources 348:193–200. https://doi.org/10.1016/j.jpowsour.2017.02.083

31. Patel K, Chikkali SH, Sivaram S. 2020. Ultrahigh molecular weight polyethylene: Catalysis, structure, properties, processing and applications. Progress in Polymer Science 109:101290. https://doi.org/10.1016/j.progpolymsci.2020.101290

32. Mader J, Xiao L, Schmidt TJ, Benicewicz BC. 2008. Polybenzimidazole/acid complexes as high-temperature membranes. Fuel Cells II 63–124. https://doi.org/10.1007/12_2007_129

33. Xu Z, Gehui L, Cao K, Guo D, Serrano J, Esker A, Liu G. 2021. Solvent-resistant self-crosslinked poly (ether imide). Macromolecules 54(7):3405–3412. https://doi.org/10.1021/acs.macromol.0c02860

34. Agopian JC, Teraube O, Charlet K, Dubois M. 2021. A review about the fluorination and oxyfluorination of carbon fibres. Journal of Fluorine Chemistry 251:109887. https://doi.org/10.1016/j.jfluchem.2021.109887

35. Fukushima M, Colombo P. 2012. Silicon carbide-based foams from direct blowing of polycarbosilane. Journal of the European Ceramic Society 32(2):503–510. https://doi.org/10.1016/j.jeurceramsoc.2011.09.009

36. Sathishkumar TP, Satheeshkumar S, Naveen J. 2014. Glass fiber-reinforced polymer composites—A review. Journal of Reinforced Plastics and Composites 33(13):1258–1275. https://doi.org/10.1177/0731684414530790

37. Ahmad F, Akhtar KS, Anam W, Mushtaq B, Rasheed A, Ahmad S, Azam F, Nawab Y. 2023. Recent developments in materials and manufacturing techniques used for sports textiles. International Journal of Polymer Science. https://doi.org/10.1155/2023/2021622

38. Basit A, Latif W, Ashraf M, Rehman A, Iqbal K, Maqsood HS, Jabbar A, Baig SA. 2019. Comparison of mechanical and thermal comfort properties of Tencel blended with regenerated fibers and cotton woven fabrics. Autex Research Journal 19(1):80–85. https://doi.org/10.1515/aut-2018-0035

39. Rathod A, Kolhatkar A. 2014. Analysis of physical characteristics of bamboo fabrics. International Journal of Research in Engineering and Technology 3(8):21–25.

40. Ławińska K, Serweta W, Jaruga I, Popovych N. 2019. Examination of selected upper shoe materials based on bamboo fabrics. Fibres & Textiles in Eastern Europe. http://dx.doi.org/10.5604/01.3001.0013.4472

41. Ryu J, Han DY, Hong D, Park S. 2022. A polymeric separator membrane with chemoresistance and high Li-ion flux for high-energy-density lithium metal batteries. Energy Storage Materials 45:941–951. https://doi.org/10.1016/j.ensm.2021.12.046

42. Kuo CJ, Lan WL. 2014. Gel spinning of synthetic polymer fibres. In Advances in Filament Yarn Spinning of Textiles and Polymers. Woodhead Publishing, 100–112. https://doi.org/10.1533/9780857099174.2.100

3 Manufacturing Technology for Functional Technical Fabrics

3.1 INTRODUCTION

Manufacturers have a diverse range of fiber materials at their disposal, allowing them to create textiles that are customized to meet specific needs and intended uses. There are four main types of fabrics: woven, nonwoven, knitted, and braided. To achieve them, various weaving techniques are used, like processes for knitwear production, nonwoven production, braiding processes, 3D weaving, braided fabrics, and composites, which are fibers in a matrix of resin, used in the automotive and aerospace industries [1]. These manufacturing techniques are employed to impart either aesthetic or functional properties, depending on the specific requirements of the end product. This category includes technical products that are woven, braided, or knitted, such as ropes, nets, and carpets (Figure 3.1a–e).

It also includes nonwoven products such as hygiene items like wipes and diapers, as well as filtration and construction industry products. The process of finishing fabrics, which may also take place earlier during the fiber or yarn finishing stages of the production process, occurs after the manufacture of woven, knitted, or nonwoven fabrics (Figure 3.2).

The creation of ready-to-wear clothing and/or garments is the last step, which entails the mass production of textile materials assembled in accordance with their intended use, whether it be for technical, home and furnishing, or apparel textiles. Textile products must meet certain specific requirements depending on their intended field of application. These requirements include aesthetic properties, physiological properties for wear (such as skin and sweat comfort management fabrics), physical properties, and resistance to chemical and biological threats such as microorganisms and pests [2–4].

DOI: 10.1201/9781003317074-3

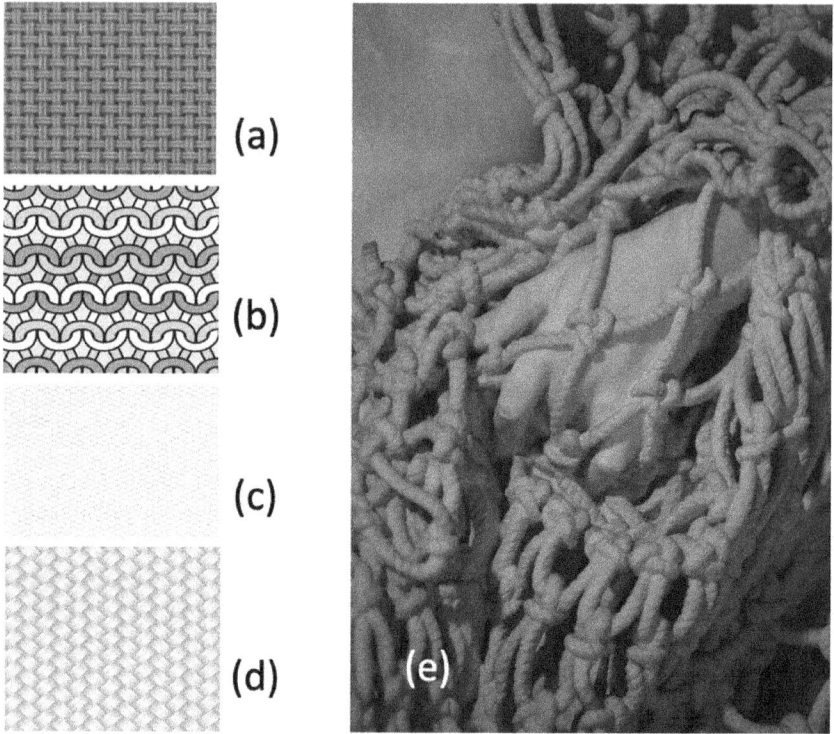

FIGURE 3.1 (a–e) Images of various techniques for manufacturing technical graded fabrics (a) woven (b), knitted, (c) nonwoven, (d) braided, (e) knotted.

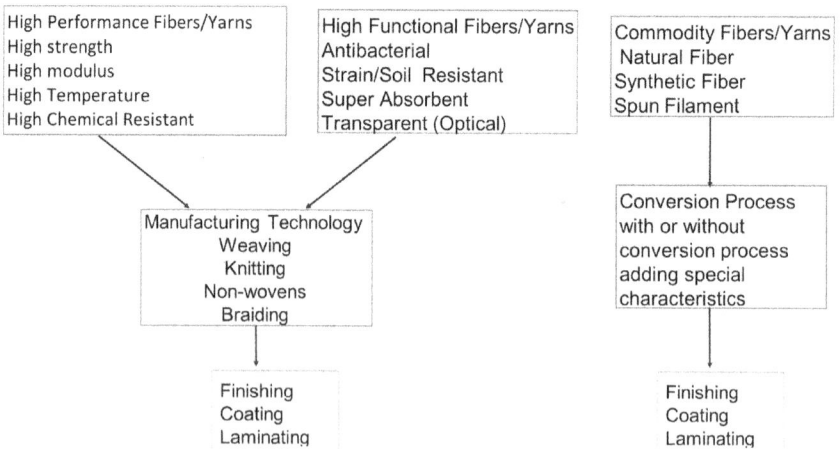

| High Performance Fibers/Yarns
High strength
High modulus
High Temperature
High Chemical Resistant | High Functional Fibers/Yarns
Antibacterial
Strain/Soil Resistant
Super Absorbent
Transparent (Optical) | Commodity Fibers/Yarns
Natural Fiber
Synthetic Fiber
Spun Filament |

Manufacturing Technology
Weaving
Knitting
Non-wovens
Braiding

Conversion Process
with or without
conversion process
adding special
characteristics

Finishing
Coating
Laminating

Finishing
Coating
Laminating

FIGURE 3.2 Manufacturing process for technical textile fabrics.

Abbreviations used in the chapter	
Abbreviation	**Full form**
EVOH	Ethylene vinyl hydroxide
FR	Fire resistant
HP	High performance
HMW	High modulus weight
HMPE	High modulus polyethylene
PBI	Polybenzimidazole
PBO	P-phenylene-2,6-benzobisoxazole
PE	Polyethylene
PET	Polyethylene terephthalate
PEEK	Polyethyletherketone
PP	Polypropylene
PPS	Polyphenylene sulfide
PTFE	Polytetrafluoroethylene
SITS	Smart interactive textile system
UHMWPE	Ultra-high-molecular-weight polyehylene
UV	Ultraviolet
VARTM	Vacuum assisted resin transfer molding

3.2   FABRIC MANUFACTURING TECHNOLOGY

3.2.1   Woven Textiles

Weaving is accomplished using various looms that differ in their speed and method for transporting filling (weft) threads. To create a woven cloth, the warp threads run longitudinally and the filling threads transversely (perpendicular to the warp). There are several factors that can impact the dyeing performance of woven textiles, including the type and amount of starch used, tension variations during weaving, and the storage of grey fabric. If starch applied on the materials become oxidized, they can become insoluble, reducing the fabric's ability to absorb dye. Tension variations during the warping and weaving processes can result in visible physical variations in the fabric after it has been relaxed and dyed. When stored improperly, the fabric can develop mildew stains that not only resist dye uptake but also contribute to fiber damage [5, 6].

3.2.1.1   Basic Weave Patterns

The warp and weft threads are crossed fundamentally and tightly in the plain weave pattern. For clothing applications, fabrics with a high thread density frequently demand the use of four, six, or more shafts. Sometimes the term "tabby" or "linen weave" is used to describe the plain weave pattern. The plain weave pattern and its fundamental modifications are shown in Figure 3.1a.

3.2.1.2   Satin Weaves

Satin weaves are known for their smooth, dense, and seamless appearance. This weaving technique is identified by crossing points that are evenly spaced apart and never meet each other.

3.2.2   SPECIAL WEAVING TECHNIQUES

Possible technical fabric structures are either woven, nonwoven, or knitted structures. These structures incorporate a number of different dyes, specialty finishes, or decorations. Some special technical graded fabric structures are shown in Figure 3.3a–e. Technical fabric can be constructed by using functional and HP base fibers, with various types of the following weave structures useful in terms of making technical fabric.

3.2.2.1   Cord and Velveteen

In the weaving process of cord and velveteen fabrics, the weft thread is floated over several warp threads and later cut. This creates a cut pile, which in cord fabric forms ribs, while in velveteen fabrics, the pile is evenly distributed.

3.2.2.2   Terry Woven

In terry weaving, the warp yarn used as the base is stretched tightly, while the warp yarn used to create the pile is left with loose tension. A set of weft threads, typically made up of three, four, five, or more threads, is inserted at a certain distance from the set of weft threads in order to achieve the appropriate pile height.

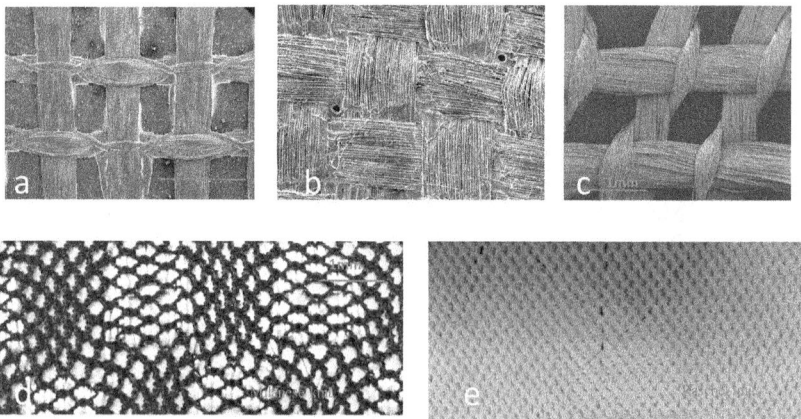

FIGURE 3.3   (a) Leno weave, (b) ripstop weave, (c) triaxial woven fabric structure, (d–e) various images of specialized fabric structures.

3.2.2.3   Two-Ply Fabric

During the manufacturing process of two-ply fabric, two separate woven fabrics are joined together by pile threads, resulting in a fabric with thickness and durability.

3.2.2.4   Matt Weave Structure

Matt weave is an extended plain weave with high energy absorption capacity and impact resistance. Matt weave fabric is suitable for technical clothing in hot climatic conditions [7].

3.2.2.5   Leno Weave

Leno weave is a sort of weaving design in which the weft threads are encircled by two or more warp threads. The tight grip of the interlaced warp threads around the weft results in a strong fabric that can withstand wear and tear (Figure 3.3a). Leno weave is characterized by a thin, open, light, and translucent appearance. Two types of warp, called doup and ground threads, cross one another alternately in the leno weave. The weave is made by crossing two or more warp threads over each other and interlacing with the weft threads at every pick (Figure 3.3a). The warp threads of the leno weave are not parallel to each other. The handloom requires a minimum of four regular harnesses, but each heald wire contains two eyes instead of one. The arrangement of the harnesses is: doup harness, standard harness, ground harness, and back harness. The alternate warp yarn crossover completes in two picks [8, 9]. Yarn interlocking makes a firm and durable structure that is not easily deformed. The high tensile strength, stretchability, and firmness of leno weave make it suitable for hybridization for medical and various customized uses [10, 11]. Medical gauze made of leno weave is used in wound dressing. The open yet stable structure of the weave is used in geotextiles, bandages, aerospace, and several technical applications [8, 9, 12].

3.2.2.6   Ripstop Weave

Ripstop weave is a special weaving technique that creates a square pattern or grid in the fabric, resulting in increased resistance to tearing and ripping. This makes it a popular choice for fabrics that will undergo coating and lamination processes, such as those used for underlining (Figure 3.3b). Ripstop weave is characterized by higher tear strength [13]. The multilayer ripstop structure protects technical textiles from harmful radiation by blocking the emissivity of electromagnetic radiation [14].

3.2.2.7   Tri-Axial Fabric

Tri-axial is the most suitable weave for technical fabric due its honeycomb structure and optical transparency/porosity. Tri-axial fabric, which has three yarn sets woven together in a single fabric to offer quasisotropic qualities and obviate the need to connect two or more plies together, allows for the transfer of stress in the bias direction. The three-yarn system's threads are oriented at a 30° angle in triaxial fabric (Figure 3.3c) [15]. This structure results in triangle intersections between the threads of the fabric, which prevent bias distortion, guarantees dimensional stability, and boosts rip strength compared to equivalent biaxial fabrics.

3.2.2.8 Laminated Fabric

The lamination of an unsupported high-modulus film to the base substrate is another material structure that can be used to improve the dimensional stability of an orthogonal foundation fabric. The film serves as a tensile member in the bias direction in this arrangement. The ply adhesion between the substrates has a significant impact on the performance of a film-reinforced orthogonal fabric.

3.2.3 KNITTED TEXTILES

Fabric is made through knitting, which is the process of interlocking or intermeshing yarn loops. Weft knitting and warp knitting are the two main types of knitting that can be done on different kinds of machines. Weft knitting involves needles knitting yarn across the width of the fabric to form loops, while warp knitting involves needles knitting a series of parallel warp threads to create loops. Knitting can be used to produce a variety of textiles, including hosiery, sweaters, and other garments, and can be done with any type of yarn or fiber [16, 17]. However, inconsistencies in tension during knitting or dyeing can cause uneven coloration, resulting in striped fabrics. A basic plain knit structure is shown in Figure 3.1b. Weft-knit and warp-knit are the additional categories into which knit fabrics can be divided. Milanese knit, a type of special warp knit fabric, is made of two sets of threads knitted diagonally, resulting in a fine vertical rib on the face side and a diagonal structure on the reverse side (Figure 3.3d). This lightweight and smooth fabric has a drapey quality and is commonly used for gloves. Raschel knit is a different kind of knit fabric that is created using a single set of latch needles and spun threads (Figure 3.3e). This type of knit can be highly patterned, lacy, or even piled and has a range of properties including density, compactness, loftiness, stability, stretchiness, and reversibility. Raschel knit fabrics can be either open or lofty and can have a single face or be reversible.

3.2.4 NONWOVEN TEXTILES

Nonwovens are becoming an increasingly popular segment in the textile industry, particularly in industrial applications. They are essentially structures made by interlocking or bonding fibers, threads, or filaments using mechanical, thermal, chemical, or solvent methods to form sheets or webs (Figure 3.1c). Nonwoven fabrics can be classified into different types, including heat bonded, wet, spun-bonded, spun lace, melt-blown (micro-denier), and hydrophilic nonwovens. Among these, spun-bonded nonwovens have diverse applications, from apparel to filters and insulation, as depicted in Figure 3.4, and a process flow diagram is shown in Figure 3.5. These materials include unique fibers like binding,

| Antibacterial | Hydrophilic | Hydrophobic | UV treated | Flame retardant |

Spun bond nonwovens

| Antistatic | Coated | Printed | Elastic | Lofty |

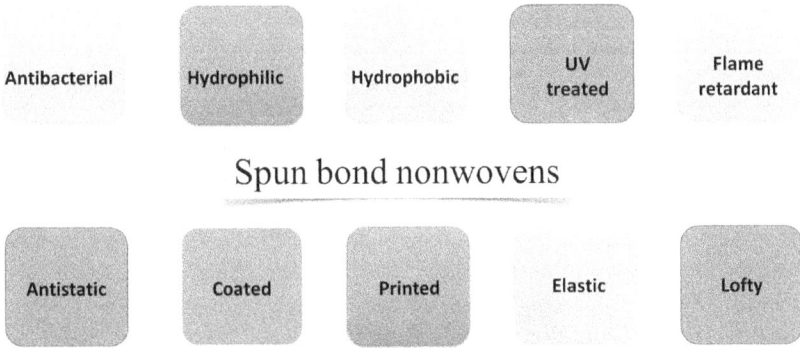

FIGURE 3.4 An application of spun bonded nonwoven fabric.

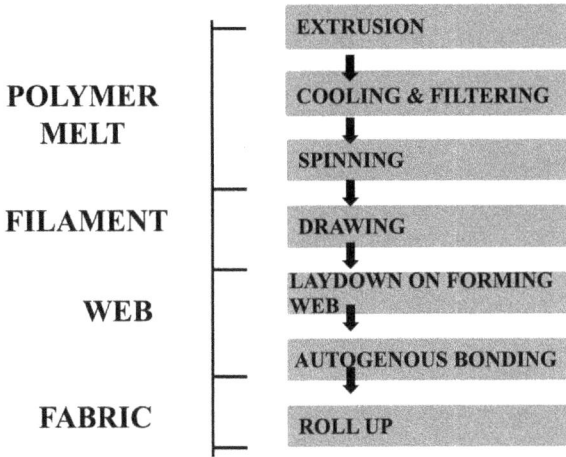

POLYMER MELT

- EXTRUSION
- COOLING & FILTERING
- SPINNING

FILAMENT

- DRAWING

WEB

- LAYDOWN ON FORMING WEB
- AUTOGENOUS BONDING

FABRIC

- ROLL UP

FIGURE 3.5 Spun bonding fabric process flowchart.

bicomponent, micro-, and shrinkage fibers in addition to human-made fibers like cellulose (for example, viscose) and synthetic polymers like polyamide, polyester, polyolefin, or fibers formed of polyacrylnitrile. It is essential to handle the calendaring process with care to ensure the fabric has the desired feel and texture [18].

Spun-bonded nonwovens offer potential for application-specific product process optimization, such as crop covers, seed blankets, weed control fabrics, greenhouse shadings, root bags, and such agricultural applications. Geotextile applications of nonwovens are in asphalt overlay, sedimentation, soil erosion, impregnation bases, drainage channel liners, separation reinforcement, roadside and drainage, and river and canal banks.

Another common nonwoven fabric used in infant diapers, training pants, feminine care items, and adult incontinence products is hydrophilic spunbonded PP nonwovens. Microdenier nonwoven fabrics, specifically melt-blown, are highly promising materials for the development of lighter-weight UV protective layers and thermal insulating layers for technical textile applications, as shown in Figure 3.4.

3.2.5 COATED AND LAMINATED FABRIC TECHNOLOGY

Enhancing the functional performance characteristics of technological textiles and consequently raising their value requires the use of coating and laminating processes. The use of these techniques is expanding quickly as a result of the expansion of technical textile applications.

However, adhesion of films with fabrics is a critical issue that can lead to poor adhesion, decreased fabric strength, susceptibility to UV photodegradation, and reduced usable life. In order to produce high-strength laminated fabrics, it is necessary to address this technology for high-end products. To achieve this objective, a proposed research project has been developed (as shown in Figure 3.6), which includes the preparation of inert and impermeable films; the development of high-strength fabric structures with different weaves, fibers, and constructions; and the lamination of films with fabrics [19].

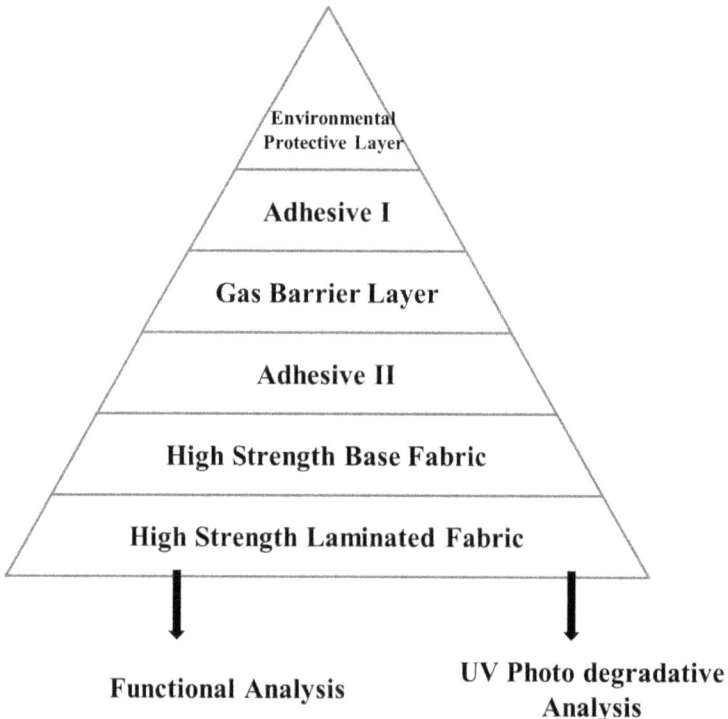

FIGURE 3.6 Process for manufacturing coated and laminated fabrics.

This study will focus on testing the inflatable materials' components for their UV performance and degradation to estimate their lifespan. This will involve conducting various R&D trials, studies, and experiments to optimize the weaving process on miniature looms, using high-performance fibers, studying film properties and adhesion behavior, and laminating the film over the fabric. Fabric development will also be an important aspect, which includes studying the effect of weave structures, fibrous materials, deniers, and triaxial fabric on mechanical properties and leakage rate. The effects of accelerated weathering on these properties will also be established.

The mechanical characteristics and fabric stiffness of the woven fabric structure, which govern the effectiveness of the coating/lamination process, will be greatly influenced by the weave floats.

Therefore, an optimized weave design will be developed to ensure a higher degree of fixation of the film onto the fabrics. To achieve the stringent norms and challenging goals, two approaches will be followed in the proposed R&D activities. The first approach involves studying various available commercial films, such as Tedlar (polyvinyl fluoride), Teflon (polyvinyl tetra fluoride), Kapton film, polytetrafluoroethylene, EVOH, and PET, to identify suitable laminated fabrics with high-strength fabrics of PET (Vectran HT) and UHMWPE (Spectra/Dyneema) with various construction deniers and weaves (ripstop, leno weave, and tri-axial). These laminated fabrics will be tested for photo-degradative performance and functional analysis for inflatable components. Lab-scale testing for UV protective properties will be carried out using a xenon arc light spectrum that is very close to the sunlight spectrum. Polypropylene and polyester films commercially available in India will also be explored for such functional textile applications. PET and PP will be coated with suitable fluorocarbon and vinylidene compound additives of various add-ons to study the photo-adhesion and functional behaviors of high-strength laminated fabrics. All these materials need to be studied in depth to understand the behavior and mechanisms of UV degradation and causes of leakage on exposure to sunlight. The generated R&D data will be utilized to establish bulk-scale fabric production. This proposal advocates for cross-disciplinary research collaboration between chemical and textile technologists from both academia and industry on a unified platform. The aim is to advance the field of technical textile materials through the application of multi-disciplinary research in this area of focus. Such an approach is more conducive for tackling complex and diversified research challenges where the starting materials and end products (i.e., fabrics) are interconnected and interdependent.

Antimicrobial or antifungicidal surfaces are useful in the design of other biocompatible, biorepulsive, or bioattractive conservation fabric surfaces. Hydrophilic and hydrophobic properties, oil repellence, soil repellence, barriers to water penetration, and antibacterial surfaces are valued for conservation fiber. Specialty coated and laminated textiles are highly beneficial for functional technical textiles to impart specialty chemicals for safeguarding and introducing functionality by lamination with functional film or fabrics layers [5].

3.2.6 SPECIALIZED MANUFACTURING TECHNOLOGIES SUCH AS TRIAXIAL WEAVE AND LENO WEAVE

Technical textiles can be produced using normal/commodity fibers or high-performance/functional fibers and threads. If normal fibers are used, functionality can be added during the conversion into fabric, at the finishing stage, or both. These textiles can introduce specialty chemicals and functionality by laminating with functional films or fabric layers as stipulated previously.

Technical textile fabrics can be created using functional and high-performance fibers, using a variety of weave structures such as matt weave, leno weave, ripstop fabric, grid structure, and tri-axial fabric. Among these, tri-axial fabric is particularly suitable for technical textile applications due to its honeycomb structure and optical transparency/porosity (Figure 3.3c).

Three thread sets are interwoven to create a triaxial fabric, which offers quasi-isotropic qualities without the requirement for many plies to be bonded together. In comparison to a comparable biaxial fabric, the three-thread system's threads are oriented at 60° or 120°, creating triangular intersections that minimize bias distortion, guarantee dimensional stability, and boost rip strength. A further technique for enhancing the dimensional stability of an orthogonal base fabric is to laminate a high-modulus, unsupported film to the substrate.

Specialty-coated and laminated textiles offer many benefits for technical fabrics, including antimicrobial or antifungicidal surfaces that can also be used in the design of biocompatible, biorepulsive, or bioattractive fiber surfaces. These textiles can have a range of surface-related properties, such as hydrophilic and hydrophobic surfaces, oil and soil repellency, water penetration barriers, and antibacterial properties. Additionally, co-extrusion of polymers and antimicrobial materials can result in fibers with antimicrobial properties. To make acetate fiber antimicrobial, an antimicrobial agent like triclosan is incorporated within the acetate fiber's interstices. Scientific research has shown that triclosan prevents the growth of a wide variety of bacteria, mold, mildew, and fungi. Since the antimicrobial agent is not water soluble, it remains embedded within the fibers, and the antimicrobial protection is designed to be long lasting and durable throughout the product's lifespan.

3.2.7 3D FABRIC TECHNOLOGY

A textile structure is often fibrous, soft, and adaptable. Due to their higher longitudinal dimension, fibers and threads are typically thought of as being one-dimensional. However, fabrics manufactured from them are thought of as being two-dimensional, and clothing made from 2D fabrics is thought of as having three dimensions. All of these materials are three-dimensional by nature, though. Based on their dimensions, woven fabrics can be divided into three types: 2D fabrics, which have their threads placed in a single plane; 2.5D fabrics, which have their threads distributed in two perpendicular planes; and 3D fabrics, which have their threads arranged in three perpendicular planes. Thus, a single fabric system made up of constituent threads arranged in three mutually perpendicular

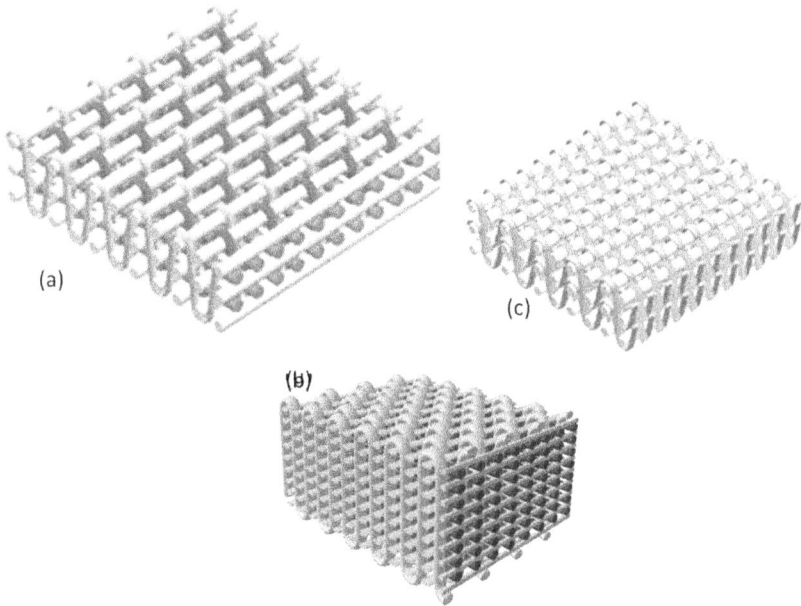

FIGURE 3.7   3D fabric structures: (a) 3D orthogonal woven, (b) 3D layer to layer angle interlock, (c) thickness angle interlock.

planes can be used to define 3D fabrics. The three main weaving techniques used in the production of woven 3D fabrics are 2D weaving, 3D weaving, and noobing [20]. A typical 2D weaving mechanism that interweaves two sets of strands produces interlaced 3D fabrics, which have significant measures in three dimensions (Figure 3.7a-c).

In contrast, weaving three different sets of threads together produces an interwoven 3D fabric. This stands in contrast to conventional textile fabrics, which possess significant measurements in only two dimensions, as demonstrated in the accompanying illustration.

3.2.8   BRAIDING

Braiding is a special process of weaving in which three or more strands or strips of yarn get twisted in diagonal fashion and overlapping to create a flat or tubular multidirectional rope structure. Braided product obtained by such technique exhibits distinct texture and appearance (Figure 3.1d).

Diamond, regular, and Hercules braids' thread interlacement topography corresponds to that of plain, 2/2 twill, and 3/3 twill woven structures, respectively.

Braid structures are classified further as two-dimensional braids, three-dimensional braids, and hybrid braids. Two-dimensional braids are biaxial (made of two sets of yarns) and triaxial (3 sets of yarns). 2D braid structures are either soutache, tubular, or flat. Three-dimensional braids are interconnected multilayered tubular or rectangular structures. A third thread is fed parallel to the axis of the braids to improve the strength and stiffness of 3D braid structures. Compared to similar 2D tubular materials, 3D composites have higher transverse strength and energy absorption capacity [21, 22]. Hybrid braids are constructed by using two or more different types of material. Traditionally braids are used as ropes, archery strings, and shoelaces. Technical applications of braids are in fiber-reinforced composites and medical implants. The advantages of a braided structure in reinforcement composites are scope for tensile compression, expansion, and improved modulus. Braided hydraulic and fuel hoses used in cars, marine engines, rocket launchers, and airplane parts have excellent bursting strength and flex fatigue [21, 23].

3.3   FOCUSED RESEARCH ON MANUFACTURING TECHNOLOGY

3.3.1   SAILCLOTH

Sailcloth is a compactly woven heavier canvas fabric. Sailcloth is made of synthetic fibers such as nylon, polyester, carbon, UHMWPE, and aramid. Bulky sailcloth is used for sailing and tents. Light to medium weight sailcloth are used in clothing accessories such as bags, hats, and in home décor. It's a weather resistant and breathable fabric.

3.3.2   COATING TEXTILES FOR HEAT MANAGEMENT

Scientists at GEMTEX are investigating the transfer of heat and moisture through multiple layers and porous coatings using both experimental and computational methods. They aim to create fibers and membranes that can adapt to changing microclimate conditions by possessing hydro- or thermos-sensitive properties that alter their air and/or moisture permeability accordingly [24].

3.3.3   MEDICAL TEXTILES AND MEMBRANE-COATED FABRICS

GEMTEX is focused on developing functional materials that can be integrated into fibers through coating or textiles. These materials are used in a variety of medical devices, including surgical hoses, hernia vascular substitutes, and blood dialysis filters. The company is particularly interested in using these materials to create minimally invasive products that support health and preserve body functions, with a special focus on meeting the needs of the elderly [25].

3.3.4   MULTIFUNCTIONAL METAL COMPOSITE COATED FABRICS FOR SMART INTERACTIVE TEXTILE SYSTEMS

A smart interactive textile system (SITS) is a technology that enables textiles to respond and interact with the environment, providing an intelligent environment for the user. To make SITS possible, textile processing techniques like spinning, weaving, knitting, embroidery, and finishing are adapted to produce conductive textiles that form the foundation of the infrastructure. Metal composite threads and fabrics are created using standard textile-processing equipment, making them practical and affordable to apply. These metal composite threads are designed to withstand specific textile processes, such as weaving, embroidery, and braiding. Depending on the required properties, metal composite fabrics such as metallic mesh woven for frequency-selective electromagnetic shielding, metal composite braids for signal and power transfer, and metal composite embroidery for radio frequency (RF) engineering may be utilized in smart textile systems [26, 27].

3.3.5 CONDUCTING POLYMERIC-COATED FILMS

There is significant interest in developing polymer films due to their ability to combine the electronic properties of metals and semiconductors with the processing advantages of polymers. Biomedical and therapeutic research committee (BTRC) is collaborating with Dundee University Dental School and has received funding through a proof-of-concept award to work on formulating, processing, and fabricating polymer electrodes for the detection of dental caries.

3.3.6 LIQUID-REPELLENT COATED TEXTILES

Stuart A. Brewer and Cohn R. Willis conducted research at the Defence Science and Technology Laboratory that demonstrated the use of specific plant features to produce surfaces that repel low surface–tension fluids, both polar and non-polar. By incorporating "hairy" fibers with fluoropolymer coatings, liquid-repellent textiles can be developed with enhanced effectiveness [28, 29].

3.3.7 LATEX-COATED TEXTILES

This US patent pertains to a novel polymer latex that can effectively absorb ultraviolet light and maintain colorfastness for extended periods when applied to various substrates such as plastics and fabrics. The unique latex is produced by copolymerizing benzotriazole or benzophenone monomers with acrylic acid co-monomers while incorporating a chain transfer agent during emulsion polymerization. The resulting latex exhibits exceptional ultraviolet absorption properties, and once applied to fabrics, it adheres well and is resistant to removal, making it a cost-effective solution for enhancing UV protection and color durability [30].

3.3.7.1 UV Protective Technology
1. According to research by Ibrahim, Allam, El Hossamy, and El Zany, a brand-new technique has been created for enhancing the UV-protective qualities of cotton/wool and viscose blended textiles, making them appropriate for premium, trans-seasonal clothing. According to the experimental findings of the study, the type of finishing chemicals

employed affects how much the UV-protection factor (UPF) of the final fabrics is improved. The UPF values are also influenced by the type of substrate. Additionally, a considerable rise in UPF values is produced by post-treatment with cu-acetate or post-dyeing using the dyestuffs that were utilized. Finally, a significant factor affecting the UPF values of the materials is the finishing regime [30].

2. Xin, Daoud, and Kong researched a new way for using the sol-gel process to create a UV blocking treatment for cotton fabrics. The process of treatment entails coating cotton fibers with a thin layer of titanium. According to Australian/New Zealand regulations, the treated fabrics achieve an outstanding level of protection against UV radiation with a UPF factor of 50+.

3. Ciba Specialty Chemicals has introduced Tinosorb FD, an innovative additive for laundry products that enhances the ability of clothing to protect against UV radiation. With Tinosorb FD, consumers can now utilize the company's advanced UV technologies to minimize their exposure to harmful sun rays simply by wearing treated clothing, providing a groundbreaking new method of sun protection.

3.3.8 Cosmeto-Fabrics and Skin Care Coated Fabrics

The use of cosmetotextiles, or clothing intended to come into touch with the skin and transmit active ingredients for cosmetic purposes, particularly in reversing the signs of ageing, is on the rise. There is a need for goods that improve beauty and prevent ageing since people in industrialized countries live longer and want to look young. Active substances like aloe vera gels, which can be gradually released and absorbed via the skin, are one popular way to accomplish this [31, 32].

1. Cognis Deutschland, based in Düsseldorf, Germany, has developed a skincare system called Wellness to Wear that can be incorporated into garments using their patented Skintex technology. Chitosan, a material produced from shrimp shells, is used to microencapsulate the active chemicals and turn ordinary clothing into "active wellness" apparel.

2. Chitosan, one of the key components in Skintex, has moisturizing properties that help to prevent skin dehydration and maintain a soft and smooth texture. The active ingredients are released through two mechanisms when wearing Skintex clothing: first, the movement and friction of the fabric against the skin, and second, the slow breakdown of the chitosan layer by the body's own enzymes, which activates the ingredients and allows them to penetrate the skin [33].

3. Cognis has unveiled a new product called Skintex Reloading, designed to address the growing need for long-lasting effects in textiles. Skintex Reloading offers the ability to recharge fabrics with new microcapsules, ensuring that they remain effective for a longer period of time. Due to the inclusion of premium skincare components, including shea butter,

apricot kernel oil, rose hip oil, and red algae extract, this cutting-edge technology is very beneficial for slimming results. Additionally, the cooling and moisturizing qualities of garments treated with Skintex can be renewed by using Skintex Reloading.

4. The Loughborough, UK-based Specialty Textile Products Limited sells the Biocap line of functional biocapsules, which include aloe vera and healthy components like vitamins A, D, and E. The wearer can receive the functional advantages of these capsules by wearing innerwear, T-shirts, stockings, socks, and bedding, among other textile products.

3.3.9 Antimicrobial and Deodorizing Coatings/Finishes

1. Thilagavathi, Bala, and Kannaian conducted research on the use of neem and Mexican daisy extracts for antimicrobial finishes on cotton fabric. They applied the extracts directly and through microencapsulation using a pad-dry-cure technique. To improve the resilience of the finish, they used phase separation/coacervation to microencapsulate the extracts using acacia as a wall material. The resulting microcapsules showed excellent resistance to microbes even after 15 washes [34].

2. Specialty Textile Products Limited markets the AR Cap, an antibacterial microcapsule containing a nonionic product with a pH of 70.5 (1% solution). At concentrations between 1% and 2%, this substance is highly soluble in liquids and efficient against a variety of bacteria.

3. A patented technology called Silver Cap has been developed, which involves placing silver nanoparticles on the outer wall of microcapsules. When sprayed on textiles, silver ions are said to kill over 650 distinct viruses and offer very strong antibacterial defense. The wall covering may be constructed of natural or synthetic polymers and may also contain additives like aloe vera, perfume, color-changing pigments, or other antibacterial substances.

3.3.10 Metallized Coating

Gentex, a company based in the United States, specializes in producing aluminized fabrics that are utilized in fire-fighting clothing to provide protection against intense radiant heat and shield industrial workers from hazardous molten metal splatters. The unique five-layer textile construction of these fabrics is capable of reflecting up to 95% of infrared heat [35].

3.3.11 Membrane Technology

Switzerland-based company Schoeller Textil has recently developed a new membrane technology called C-change, which has the ability to adjust its moisture vapor permeability according to different weather conditions. This innovative "bionic climate membrane" is both water and windproof, and it can open its structure to allow excess heat to escape when temperatures are high or when physical activity is intense. Conversely, in cold weather or during periods of low activity, the membrane's structure closes to retain body heat.

3.3.12 BARRIERS AND ANTIMICROBIALS

Skymark, a UK-based company, has developed Skyair membranes using advanced resin and manufacturing technology, which are waterproof, windproof, and breathable. These monolithic membranes have no pores, making them an excellent bacterial barrier while also allowing for breathability. As a result, they are suitable for use in both outdoor clothing and healthcare settings, where they can protect staff from bacteria, viruses, blood, and other fluids while providing comfort. The Devan Group, based in Belgium, has created the Aegis Enhanced range of antimicrobial chemical treatments, which can enhance comfort and well-being. This range includes various chemical finishes that provide moisture management, stretch recovery, sensory management, shrink resistance, flame retardancy, and anti-static technology, all of which are also antimicrobial-A multifunctional fabrics [36, 37].

An antimicrobial fiber based on acetate has been developed, as described in a US patent study, and is sold under the name Microsafe by Hoescht Celanese of South Carolina. The acetate fiber is a type of manufactured cellulosic fiber produced from wood pulp and acetic acid. To impart antimicrobial properties to the fiber, an antimicrobial agent like triclosan is embedded within its interstices. Triclosan has been scientifically proved to stop the development of certain bacteria, mold, mildew, and fungus and remains embedded within the fibers since it is not water soluble. Additionally, the antimicrobial protection is designed to be long-lasting and durable, allowing the fiber to provide continuous antimicrobial protection throughout the product's lifespan. The cellulosic nature of the fiber allows for breathability and absorbency, but its strength may not be sufficient for cleaning applications. To increase the strength of the acetate antimicrobial fibers, they are typically blended with synthetic fibers such as rayon, nylon, or polyester, with a preferred denier range of 250 to 450. To address the issue of deterioration of antimicrobial and deodorizing properties during use, Japanese researchers have developed an antimicrobial textile product that includes acrylic fibers with a prescribed amount of antimicrobial agent, such as chitosan, in a specific mixing ratio. This product is resistant to washing and does not cause discoloration, maintaining a balance between antimicrobial effect and pH level while being safe for humans and other organisms. Additionally, the functionality of the fiber can be easily modified during the spinning process, which is not possible with regular commodity fibers [19, 38, 39].

3.4   SPECIAL CLASSES OF TECHNICAL TEXTILE MATERIALS

The future is set to revolve around smart and functional products that incorporate artificial intelligence. Every industry is striving to create such products, and the textile industry is no exception. In order to keep up with technological advancements, the textile industry is also focusing on developing intelligent and functional products.

3.4.1 PHASE-CHANGE POLYMERIC MATERIALS

Phase-change materials (PCMs), often referred to as latent heat-storage materials, are frequently employed in the creation of heat-storage and thermos-regulated textiles and garments because of their thermoregulating capabilities. As their temperature rises consistently during the heating process, these materials can absorb heat, storing it within the material. The heat that has been stored is later released into the environment via a reverse cooling procedure. The most popular kind of PCM can switch between a solid and a liquid state, then back to a solid one. When the PCM melts, heat is absorbed between 20 and 40 °C, and when it crystallizes, heat is released between 30 and 10 °C. Paraffin waxes are the most frequently utilized type of smart phase-change materials, which can be divided into numerous types. The use of liquid-to-gas phase-change materials and non-paraffin organics as alternatives to conventional PCM materials for electronics heat sinks is uncommon. On the other hand, metallic PCMs are useful in high-temperature situations where adequate paraffin waxes are hard to obtain. A number of compounds, including hydrated inorganic salts, polyhydric alcohol-water solutions, polyethylene glycol, polytetramethylene glycol, and aliphatic polyester, have restricted temperature ranges that correlate to the phase transition temperature of those materials. These materials can also be utilized for controlled release of microencapsulated fragrances, vitamins, and other substances in new brands of clothing. Overall, PCM-based textiles provide an effective means of thermoregulation and heat storage.

3.4.2 PHASE-CHANGE PRODUCTS

BASF SE, Crypac, Dupont De Nemour, MicroCaps solvenia, PLUSS are few of the phase-change products (PCP) making companies that specializes in phase-change materials and related products. It offers a range of phase-change fabrics designed for different industries, such as medical, construction, and textiles. It's important to note that the availability and specific brands of phase-change fabrics may vary over time as new manufacturers and products emerge in the market.

There are several manufacturers that produce phase-change fabric brands. One notable manufacturers is Outlast Technologies, a leading manufacturer of phase-change materials and fabrics. It offers a range of products including Outlast Adaptive Comfort fabric, which utilizes phase-change technology to regulate temperature and improve comfort.

3.4.2.1 Coolcore

Coolcore is another well-known manufacturer of phase-change fabric. It specializes in developing cooling fabrics that use phase-change technology to manage heat and moisture, providing enhanced comfort for various applications.

3.4.2.2 Schoeller Technologies

Schoeller Technologies is a company that offers innovative textile solutions, including phase-change fabrics. Its PCM fabrics provide temperature regulation and can be found in various performance apparel and bedding products.

3.4.2.3 Nanotech Energy

Nanotech Energy is a manufacturer that focuses on advanced materials, including phase-change fabrics. It develops PCM-infused textiles that offer thermal management properties and can be used in diverse applications.

3.4.3 ELECTROACTIVE MATERIALS

The field of technical textiles has developed methods to create fabrics that are conductive and can be used for electronic wearable textiles. Traditional cotton, polyester, and nylon are used as well as more modern, practical Kevlar materials. There are two main methods employed: (1) integrating electronic sensors or devices into the woven fabric and (2) printing with electroactive polymers such as polythiophene or polypyrrole to create a continuous electrical circuit. Wearable electronics for smart electronic fabrics (e-textiles) are created using these electroactive polymer-based sensors, actuators, electronic components, and power sources.

However, further research is necessary to fully integrate actuating and sensing elements into a woven device. An innovative approach involves using standard industrial processes to create the active and sensing elements within the same woven system. Results have shown that electroactive functions can be implemented alongside vital signs and movements, creating readable signals that can be acquired and transmitted to elaboration devices. Due to this, multifunctional wearable human interfaces have been created. These devices are valuable tools in a number of biomedical sectors, including telemedicine, rehabilitation, and biomonitoring. These materials may be used in smart electronic wearable textiles in addition to sensing, actuation, electronics, and energy generation/storage. E-smart wearable textiles have the potential to revolutionize soldiers' uniforms by integrating sensors and charging stations. This would enable soldiers to travel through challenging terrain with less weight to carry.

3.4.4 SMART COLORANTS (CHAMELEON)

Functional and smart dyes can be found in both synthetic and natural forms, but due to environmental concerns, researchers are exploring natural sources for these dyes. While functional dyes have primarily been used in the textiles industry, smart dyes, which offer reversible color changes, are becoming increasingly popular. Functional dyes, such as UV-protective, antimicrobial, and moth-repellent dyes, provide specific functions to textiles after application. On the other hand, smart dyes, including photochromic, thermochromic, electrochromic, and solvatochromic dyes, are being used to create smart textiles that can change color reversibly, leading to the development of various properties like thermoregulation and camouflage. In camouflage fabric, the color of the dye can be reversed, and the reflectance can be matched to the terrain, making personnel and equipment invisible to adversaries. Chameleon high-performance fibers, extruded with smart coolants, can conceal persons or equipment, but research is needed to determine wash and rubbing fastness and fiber affinity of dyes on various textile substrates. Smart dyes are being used to create textiles with reversible color changes, which can also be applied in various fields like automobiles, robotics, aircrafts, medicine, and surgery. In summary, natural sources of functional and smart dyes are being explored to address environmental concerns.

3.4.5 FUNCTIONAL GRADE MICROPOROUS MEMBRANES AND FILMS

Commercially available films such as polyester, polyurethane, fluorocarbon, PTFE, PVF, PVDF, acrylics, and polyamino acids are used in the manufacturing of multi-layered structures through lamination techniques with high-performance fabrics. The uniqueness of these films lies in their impermeability to gases while still being permeable to water vapor and air droplets. These functional films are employed in layered structures for air balloons and extreme cold weather clothing.

3.4.6 HIGH-PERFORMANCE CHEMICALS AND NOVEL-GRADE NANOPARTICLES

Table 3.1 provides an overview of newly developed chemicals and nanomaterials that can be used to produce functional textiles with a wide range of applications.

TABLE 3.1
The Applications of Performance Chemicals in the Technical Textile Segment

Coating Chemical	Applications
Polyvinyl chloride	Tarpaulins, coverings, large tents and architectural uses, seat upholstery, PVC polyester-tent covers, leather protective clothing, aprons, leisure products, banners, bunting
Polyvinylidene chloride	Blends with acrylics to improve FR in coating
Polyurethane	Waterproof protective clothing, waterproof/breathable protective coating, aircraft life jackets, adhesives, filters used as lacquers for PVC tarpaulins and leather
Acrylic	Back coating for upholstery, including auto seats, binders for nonwovens and glass fibers, adhesives, used as lacquers for tarpaulins
Ethylene vinyl acetate	Backings for carpets and upholstery, wall coverings, exhibition board backing, adhesives
Polyolefin, LDPE, HDPE, polypropylene	Lightweight coverings, tarpaulins (alternative to PVC), sacks, bulk bags

(Continued)

TABLE 3.1 (Continued)

Coating Chemical	Applications
Silicone	Air bags, food, medical applications, gaskets, seals, parachutes, woven curtains
Poly tetra fluoro ethylene	Architectural applications (high abrasion resistance), calendar belts, food & medical uses, gaskets, seats
Natural rubber	Carpet backing, tires, life rafts, conveyor belts, protective clothing, escape chutes
Styrene butadiene rubber	Carpet backing (as natural rubber)
Nitrile rubber (acrylonitrile/ butadiene)	Oil-resistant clothing, oil seals, belts, & items handling oily or greasy products
Neoprene rubber	Resistance to chemicals, ozone, hydrocarbon, selected oils, solvents, neoprene/glass high-temperature applications, neoprene/polyester life rafts, shelters, neoprene/nylon industrial covers, bellows
EPDM rubber	Electric insulation
Butyl rubber	Items containing gases, such as air cushions, pneumatic springs, bellows (gas barrier), protective clothing— especially for chemicals and acids, lightweight life jackets, life rafts, protective clothing, aircraft carpet backing, aircraft slide/rafts
Polychlororpene rubber (e.g. Neoprene—DuPont) CR	Hovercraft skirts, flexible gangway bellows (trains), air springs, random covers, airbags, V-belts
Chlorosulphonated polyethylene/ rubber (e.g. Hypalon—DuPont) CSM	Similar to neoprene, used where coloration is necessary and higher temperature resistance is required. Liners and covers for portable waste reservoirs and waste containment ponds. Single-ply roofing, Hypalon coated glass fabric used for thermal insulation, Hypalon coated polyester fabric for diaphragm, Hypalon coated nylon fabric used for roofing, Hypalon coated Kevlar fabric for flame retardant protective clothing, and Hypalon mixed with neoprene and coated on nylon fabric used for inflatable boats.
Fluoro elastomer (Viton—DuPont) FKM	Color resistance, low-maintenance finishes, excellent abrasion resistance, seals substrate from moisture resistance
Modified vinyl acetate	Tapes
Nitrile	Belting, tapes, gaskets
Styrene/acrylic	Building products
Nanograde metal oxide	Photocatalytic activity
Nano TiO_2	Electrical conductivity
Nano Al_2O_3	UV blocking properties
Nano ZnO	Photo oxidizing
Nano MgO	Chemical and biological species
Nano Ag	Anti-microbial
Poly(3,4-ethylenedioxythiophene)	Conductive textiles

3.5 INNOVATIVE FABRICS

This section briefly presents functional fabrics and their corresponding physical properties. Test methods for evaluating their functionality are presented in Section 3.7.

3.5.1 MELT-BLOWN NONWOVEN FABRICS

Melt blowing (MB) technology is a rapidly growing area within the field of nonwoven technologies. It offers a distinct cost advantage over other systems because it allows for the direct formation of a molten polymer web that hasn't undergone any intentional stretching. The MB process involves using high-velocity air or another force to attenuate the filaments and produce fibrous webs or articles directly from polymers or resins. One of the unique features of the MB process is its ability to produce microfibers, which are much finer than normal textile fibers. The resulting nonwoven fabric from MB is highly porous, with a higher number of smaller pores in the range of 10 to 15 μm. This porosity makes the fabric efficient in trapping cold air, thereby enhancing its insulating properties. To convert the web or sheet of fibers or filaments into fabric, various techniques are employed, including thermal bonding, chemical spraying, hydro entanglement, spun bonding, and melt blowing. Melt-blown fibers are typically ultrafine, with filament diameters ranging from ~0.1–1.0 μm. The fineness of the fibers and the small pore sizes make melt-blown technology unique compared to other nonwoven technologies.

The polymer is fed into a die tip during the melt blowing process, and the resulting fiber is attenuated by hot air that is directed close to the die tip. The fiber bundle moves forward and backward due to the interaction of ambient air and high-speed air.

As the filaments fly through the ambient air, they are stretched for the first time, and during their descent onto the belt, they are stretched again due to the "form drag" caused by changes in fiber direction. As a result, melt-blown fibers usually have varying diameters along their length.

Entanglement and cohesive sticking work together to hold the fibers in the melt-blown web together. A wide range of product features are available with melt-blown webs, such as random fiber orientation, extremely fine fibers, and low to moderate web strength. Filtration media, such as surgical mask filters, liquid and gaseous filtration, disposable surgical gowns, sterilization wrap, and disposable absorbent items like oil sorbents are where monolithic melt-blown materials are most commonly used. The uncombined (monolithic) state of melt-blown material is utilized to a degree of about 40%. Due to their superior barrier qualities and mechanical robustness, laminated structures, also known as spunbond-meltblown-spunbond (SMS), are particularly well suited for gradient filtration. Electrostatic charges are often applied to the filaments to increase filtration. In summary, melt blowing technology produces superabsorbent and strong fibers, along with absorbent and transparent webs suitable for technical textile fabrics.

Its applications range from filtration to disposable products, and it offers unique advantages such as fine fibers, fine pore sizes, and the ability to produce microfibers directly from molten polymer.

3.5.2 Typical Moisture Management Fabrics for Comfort

The concept of comfort encompasses a state of pleasant harmony between individuals and their surroundings, including psychological, physiological, and physical aspects. Various types of comfort exist, including physical, neuro-physiological, and thermos-physiological comfort. Thermo-physiological comfort specifically focuses on maintaining the human body at a core temperature of 37 °C.

In normal conditions of human activity and environmental factors, the heat generated by metabolism is transferred through conduction, convection, and radiation. To regulate body heat, the body perspires in the form of vapor. However, in colder ambient temperatures, there is a higher temperature gradient between the human body and the environment. As a result, the rate of heat transfer increases, causing the body temperature to decrease rapidly. To counteract this, it becomes necessary to wear multiple layers of fabrics for protection.

A multi-layered moisture management fabric is meant for extreme cold climate as an enhanced comfort property. Cold weather clothing typically consists of three functional layers: due to its close proximity to the skin, the base layer needs to have high vapor permeability in order to effectively drain away perspiration and moisture from the body surface and transmit it to the following layer. Inner layers often utilize materials with high wicking properties, such as silk and polypropylene fabrics. The middle layer's primary function is to provide high insulation. Materials used in this layer are chosen for their ability to trap air, which aids in insulation. Wool and polar fleece fabrics are commonly used in this layer, and nonwoven fabrics may also be utilized. In terms of heat transmission, conduction is particularly important for this layer. The outermost layer serves as a protective barrier and needs to be both waterproof and breathable. It prevents cold air and water from reaching the inner layers of the fabric. Additionally, the fabric should have a high moisture vapor transmission rate to prevent dampness, which can reduce clothing insulation. The comfort properties of these multi-layered fabrics are analyzed to enhance the functionality and effectiveness of clothing designed for extreme cold climates.

3.5.3 Breathable Coated Fabrics

As a result of the presence of polar groups, hydrophilic polymer-based breathable coatings have high water vapor permeability. When these polymers are utilized on textile substrates, these groups also cause the polymers to be water soluble and susceptible to poor wash fastness. In order to integrate hydrophilic polymers into textile substrates and get over this restriction, an appropriate cross-linker is needed. These polymer-coated textile substrates are used for a variety of hazardous environments, including protective fabrics, survival suits, diving suits, mountaineering suits, specialized military clothing, chemical splash suits, surgical gowns, prosthetic socks, rainwear, cold weather jackets, tents, transport cargo wraps, sleeping bag covers, and more. A process that involves absorption of water vapor at the surface with higher concentration, diffusion through the film controlled by the concentration gradient, and desorption at the surface with lower concentration is made possible by solid hydrophilic polymer coatings that are free of pinholes, microcracks, and other flaws. Hydrophilic coatings are typically not very hydrophilic in nature. They are made of copolymers with hydrophilic and hydrophobic segments, where the hydrophobic moieties maintain the integrity of the textile substrate and the hydrophilic segments permit the passage of water vapor. However, the passage of water vapor is constrained by such coatings. Highly hydrophilic materials are thought to be the most advantageous in this situation. The polymers in question, including poly(vinyl alcohol), poly(ethylene oxide), and poly(acrylamide), are wholly soluble in water. As a result, they require chemical bonding to the textile substrate in order to stay integrated rather than being simply coated. Recent studies have looked at PEG-based coatings and hydrophilic breathable coatings made of polyacrylamide. Citric acid was used as a cross-linker and sodium hypophosphite as a reaction catalyst during the chemical integration of polyacrylamide onto cotton fabric, and the results were highly encouraging in terms of breathability performance compared to uncoated cotton fabric.

3.6 APPLICATION OF PLASMA TECHNOLOGY IN TECHNICAL TEXTILES

Plasma technologies provide an eco-friendly and versatile method for treating textile materials by modifying the surface of fibers. This surface modification enhances various properties such as wettability, liquid repellence, dyeability, and coating adhesion by functionalizing the textile substrate under an inert or gaseous atmosphere. There are various ways to achieve this surface modification, including

1. Plasma cleaning and etching, which includes clearing the exposed surface of contaminants or substrate material,
2. Utilizing the appropriate active monomers to add new functional groups to the surface that has been treated. The monomers subsequently engage in standard free radical polymerization on the plasma's active surface.
3. Applying atmospheric pressure plasma to cotton gray fabric that has been sized using a standard sizing formula that contains starch, air, and a mixture of helium-heated gases, which modifies the surface morphology, has a desizing effect, and improves wettability and wicking action.
4. O_2 plasma treatment, which greatly raises the pace at which cotton fabrics drain away moisture, increasing their absorbency and, in turn, their dyeing rate. Due to the ablation effect of non-polymerizing reactive plasma gas, visible holes were seen on the surfaces of cotton fabrics

treated with O_2 plasma. A cutting-edge method to improve wet textile processing, from pretreatments to finishing, is plasma processing, a dry chemical treatment technology. Plasma processing has the ability to significantly lessen the environmental impact of conventional textile processing techniques, which use a lot of water and energy; produce a lot of effluents with a lot of chemical oxygen demand; and contribute to excessive color, pH, and toxicity.

5. The major sources of effluent pollution in textile processing include desizing, dyeing, washing, and finishing [40]. Plasma processing is considered a sustainable and eco-friendly alternative for textile processing. However, the major challenges in commercializing this technology for bulk products are controlling the strength losses of substrates and competing with the cost of traditional water-based processes.

3.7   TEXTILE TESTING METHODS

3.7.1   Fabric Sett

According to ASTM D3775–03, a counting glass is used to measure the fabric sett. The ASTM D1059 standard is followed in determining the yarn linear density and fabric weight per unit area. The ASTM D1777–96 (2002) standard test method for the thickness of textile materials is used to measure the thickness of the textile material. A thickness tester with a pressure of 20 gf/cm^2 and an accuracy of 0.01 mm is used to test the thickness. Each fabric sample receives an average of ten readings for each test. Standard atmospheric conditions are kept throughout the experiment sample.

3.7.2   Resistance to Water Penetration

AATCC Test Method 127–1977 is used on a Shirley hydrostatic-head tester to assess control and coated textile water resistance to hydrostatic head. Applying hydrostatic head to the coated side of the fabric will allow the tester to compare the water permeability of the control and coated fabrics. When the first three drops of water appear on the opposite side of the fabric while applying pressure, the readings are noted.

3.7.3   Moisture Vapor Transmission Rate

The moisture vapor transmission rate (MVTR) tester (Model CS-141) is a tool for calculating the T value, which is the mass of moisture vapor, measured in grams, that flows through a 645-cm^2 patch of fabric in 24 hours when the relative humidity (RH) is at 100% and the ambient temperature remains constant. A fabric sample with a diameter of 108 mm is cut for testing, and an average of five readings is obtained for each fabric quality to ensure reliable results. The ability of fabrics to transfer water vapor is a key quality that can be measured more quickly and easily with the help of this tester. It works by continuously detecting the humidity produced in controlled environments. This measurement is based on the use of the ideal gas law and the gas permeability equation. According to the gas permeability equation, the pressure differential across the barrier is the reciprocal of the barrier's thickness and the permeability of the barrier. In this example, the fabrics are all intimately correlated with the rate of mass transfer. The moisture vapor transmission tester can determine the T value and provide important details regarding the rate of moisture vapor transmission through the fabric by using this equation and the ideal gas law.

Utilizing this tester, the following calculations are used to determine the MVTR:

Cut a cloth sample with a 10.8 cm diameter. For every fabric quality, take five measurements of the moisture vapor transmission. By comparing successive % RH measurements and averaging these differences, you can determine the average difference in % RH values (% RH).

Calculate the difference in minutes between each reading. Obtain the value of H, which represents the grams of H_2O/m^3 of air at the cell temperature. You can find the appropriate H values in the absolute humidity table of the *Langes Handbook of Chemistry* (1999). Plug the values of $\Delta\%$ RH, time interval, and H into the following formula to calculate the moisture vapor transmission rate (T):

$$T = \frac{(269 \times 10 - 7) \times \Delta\%RH \times 1440 \times H}{(Time\,interval)}$$

These calculations will give you T, which represents the fabric's moisture vapor transmission rate under the given circumstances, where the average difference in relative humidity values (% RH) is calculated using the grams of H_2O/m^3 of air at the cell temperature (H) values obtained from *Lange's Handbook of Chemistry* (1999)'s absolute humidity table and the time interval between successive readings (in minutes).

3.7.4 WATER VAPOR PERMEABILITY

According to British Standards BS 7209:1990, the water-vapor transmission rate of fabric samples, both coated and uncoated, is evaluated. An environmental test chamber (ETC) made by International Equipments, Mumbai, was used for the study. This ETC was created specifically to ensure accurate relative humidity management to within 1% and temperature control to within 0.5 °C. The chamber was 460 by 492 by 606 mm in size. A 402-mm-diameter turntable was positioned within the chamber and turned at a rate of 6 revolutions per minute by means of a microcontroller. Along the edge of the turntable, water-filled cups with a mouth surface area of 2873 mm² were placed. The fabric test samples were placed over the cup lips. Adhesive tape was used to bind the samples in such a way that there would be a 10-mm air gap between the water surface and the samples. PVA-coated samples and one uncoated control cloth were evaluated in each experiment. The studies were carried out at a constant temperature of 30 °C and relative humidity of 40%. The fabric samples affixed to the beakers were treated for 10 hours at the predetermined test conditions prior to the test.

At the start of the experiment, the beakers were weighed. The samples were then examined for a specified amount of time, expressed as t hours, at the pre-determined temperature and relative humidity. These observations were used to compute the water-vapor transmission rate (WVTR).

$$WVTR\left(gm^{-2}\,24\,hr^{-1}\right) = Weight\;loss\;of\;water\;in\;time\left(t\right) \times 106 \times 24\,/\,t \times 3000$$

3.7.5   GUARDED SWEATING HOT PLATE

Thermal resistance (insulation) and resistance to moisture vapor are both tested in the guarded sweating hot plate test. The fabric sample is placed on the guarded test plate, which is heated under regulated circumstances to a constant temperature of 35 °C, including an ambient temperature of 25 °C, a relative humidity of 65%, and an air velocity of 10.05 m/s.

The following equation governs the fabric's thermal resistance (Rth):

$$Rth = \frac{\left(Ts - Ta\right)}{\left(\dfrac{Q}{A}\right)}$$

Rth is the fabric's and the air layer's thermal resistance (m² °C/W).

The test plate's temperature (Ts), test section area (A) in square meters, and ambient air temperature (Ta) in degrees Celsius all affect how much electricity is needed to maintain the test plate's temperature (Q).

3.7.6   COATING INTEGRITY USING HARSH WASH TEST

The determination of fastness to washing is conducted following the test method outlined in the International Standards Organizations. The test involves subjecting a specimen to mechanical agitation in a launder meter, utilizing a 5-gram-per-liter (gpl) non-ionic detergent and 2 gpl soda ash. Additionally, ten steel balls are included in the washing process. Forty-five minutes of washing are spent at a temperature of 60 °C. The samples are weighed both before and after the washing cycle. The weight loss percentage is then calculated by comparing the initial and final weights. These readings are subsequently compared with the results obtained from the mild wash test.

$$\text{Weight loss}\% = \frac{\text{Initial weight - Final weight} \times 1000}{\text{Initial weight}}$$

3.7.7 EVALUATION OF WATER RETENTION

The water retention evaluation involves conducting tests on carefully selected samples. These samples are chosen to study the impact of the PVA to PAAc molar ratio at a specific coating add-on level, as well as to observe the effect of different coating add-on levels at a particular molar ratios on water retention.

To begin the evaluation, both the coated and uncoated samples were weighed before the test. Subsequently, the samples were immersed in water at room temperature for a duration of 15 minutes, allowing sufficient water uptake to occur. Once the soaking period was complete, the soaked samples were hung inside a centrifuge. Care was taken to ensure that the samples did not touch the bottom of the poly(propylene) tube. The other end of the hook used to suspend the samples was tied securely at the top of the tube.

$$\text{Water retention }\% = \frac{\left(\text{Weight of wet coated fabric after centrifuge} - \text{Initial weight of dry coated fabric}\right) \times 100}{\text{Initial weight of dry coated fabric}}$$

3.7.8 STIFFNESS (COATED FABRICS)

The criteria established by the British Standards: 3356:1961, "Determination of Stiffness", is used to evaluate the stiffness of coated fabrics. The bending length of the cloth, a measure of stiffness, is measured using the Shirley stiffness tester. To achieve consistent results, the fabrics are treated overnight before testing. The coated substrate is made by cutting samples along the warp and weft directions in accordance with the testing requirements. The testing is primarily focused on washed fabrics.

3.8 CASE STUDY SERIES

3.8.1 A SPECIAL INHERENTLY FIRE RESISTANT DEVELOPED POLYESTER FABRIC

3.8.1.1 Multifunctional Polyester Fabric

The development of an inherently FR, oil and water repellent, and antistatic polyester fabric has yielded several positive outcomes, paving the way for further growth and potential advancements in this versatile fabric. By maximizing the potential of polyester material, targeted features have been achieved. The outer fabric is primarily composed of aramids and viscose, with potential modifications to enhance their properties. One noteworthy accomplishment is the increase in breaking strength, which is 10% higher than the current values (80N-warp and 40N-weft) in both the warp and weft directions. Similar to this, the existing values of 3.2 Kgf-warp and 2.3 Kgf-weft are exceeded, increasing tear strength by 10–15% in both the warp and weft directions.

The inherent flame retardancy of aramid is leveraged to further enhance the fabric's flame-retardant properties. Additionally, the moisture absorbency of viscose is harnessed to improve the fabric's antistatic capabilities, making it suitable for apparel-grade applications and radiological dust repulsion. The denser fabric structure achieved through the combination of aramid and viscose provides superior oil and water repellency compared to conventional fabrics. Moreover, the moisture absorbency of viscose significantly enhances physiological comfort, surpassing that of polyester-based outer fabrics. Overall, the development of this multifunctional fabric has yielded significant improvements in strength, tear resistance, flame retardancy, antistatic properties, and oil/water repellency while also offering enhanced comfort through the improved moisture absorbency provided by viscose.

3.8.2   ELECTROSPUN NANOCOMPOSITES

Traditional materials like metal alloys, ceramics, and polymers frequently cannot meet the demands of modern technologies, which frequently need materials with unusual combinations of attributes, particularly in the case of applications such as aerospace, defense, underwater, and automobiles, where high specific strength, light weight, and lower energy are needed. Composite materials, particularly polymer composites, have proved useful in these areas. However, composite materials often fail in three areas: matrix cracking, fiber breaking, and delamination. Delamination is the most significant failure mode, and it can be prevented by incorporating secondary reinforcement in the form of nanocomponents, nanofibers, nanoparticles, or nanotubes. Nanofibers are often used as secondary reinforcement due to the ease and affordability of producing nanocomponents in large quantities using the electrospinning process.

This method has proved one of the most economical ways to create nanofibers with a mass ranging from 10 to 1000 nm. By raising the weight percentage of the fiber during the manufacturing process, composites can be made stronger. Strength rises along with fiber content. Vacuum-assisted resin transfer molding (VARTM) and autoclave procedures have a comparable range of 60–65% when compared to processes based on fiber weight fraction.

However, considering cost, VARTM is a more economical option than autoclave, making it the preferred manufacturing process for composites. In this study, a nanocomposite was manufactured using 7781 glass fibers as primary reinforcement and electrospun nanofibers of nylon 6 (55-nm diameter) as secondary reinforcement. The epoxy matrix used was Epolam 5051 with 30% hardener, and 5% nylon 6 nanofibers were added as secondary reinforcement. Test specimens were prepared per ASTM standard D7264 for flexural strength and were cut using water jet machining. The results of the study showed an improvement in flexural strength by 12%. By adding 5% nylon 6 nanofibers as secondary reinforcement, the glass fiber epoxy composite's flexural strength was improved [33, 41–43].

3.8.3 Used Tire Textile Fiber Weaving and Installations

Every year around 1000 million tires are disposed of, and this number is estimated to escalate to 1200 million tires/year by 2030. Used tire dumping in landfills is avoided, as they are non-biodegradable and large in size and number. Since tires contain synthetic rubber, recycling them is difficult and environmentally harmful. Stacks of abandoned tires contaminate ground water and aquaculture, serve as a mosquito breeding ground, and constitute a constant fire hazard. Natural rubber, synthetic rubber, carbon black, and additives (textile, steel, curing agents) are the main ingredients of tires. Ninety-one percent of used tires in European nations are recycled, mostly by being chipped into materials (56.4%) and used for energy (34.9%) [44–46].

An innovative method of utilizing waste tires is to make swings, planters, patio chairs, theme park components, sculptures, climbing nets, retaining walls, mulch rings, and landscape decor [47].

The scrap tire is typically converted by specialized cutters into rubber chips (20–50 mm), rubber granulate (0.8–20 mm), and rubber dust (8 mm). Shredded tire granules are used in tire reconstruction, road construction, and playground fields [48]. Recycled tire uses in building construction are in flooring, concrete blocks, railroad crossings, carpet lining, foot mats, highway barriers, and tire reefs. The practice has huge environmental and landscape design benefits. Petroleum spills are hazardous to marine ecosystems, destroying ocean organisms. Recycled rubber aerogels from waste tires are a promising material in oil spill cleaning. Aerogels are extremely porous (92.3–98.3%) and low in density [49].

3.8.3.1 West Coast Rubber Recycling

A US-based company, West Coast Rubber Recycling (WCRR), formerly engaged in collecting tires for landfills, worked on recycling tires for particle board flooring installations in public parks. Rubber tire tiles in residential parks, walkways, and sports arenas offer muscle strain resistance, control injury, and prevent soil erosion and dust. One thousand and 6000 tires are rescued from landfill areas if utilized for backyard play area and horse arenas, respectively. WCRR now operates in every part of the US either directly through installations or by supplying waste tires across the country.

3.8.3.2 De'Dzines

De'Dzines is an Indian startup working towards the sustainable recycling of used tires into utility products and installation designs. The tire is unraveled to procure long constituent threads to weave outdoor furniture. Tires are shredded to fabricate planters. Tire products include bags, planters, swings, woven chairs (Figure 3.8), and a beautiful animal park from scrap tires. Almost 2 tons of tires were used in designing the park in Kanpur, India (Figure 3.9).

FIGURE 3.8 Used tire products—fibers extracted from tires were utilized to weave outdoor furniture; tire benches—two tires were placed on the ground to provide support to the sitting bench; planters—tires were torn and molded to form the curves of the planters.

FIGURE 3.9 Used tire sculptures of lion, elephant, monkey, and horse in animal park.

3.9 CONCLUSION

To achieve the objective of developing high-tech technical textiles, the following key areas need to be established: exploration and development of special fibers, fabrics, threads, and finishing technology. To accomplish these objectives, the following tasks have been defined:

- Spinning various polymers into homofibers, composite fibers, and bi-component fibers using different fiber spinning techniques at the laboratory scale level.

- Controlling fiber morphology and diameter by selecting appropriate solution properties and operating parameters.
- Converting spun fibers into yarn and subsequently into fabric, either in the form of nonwoven or woven or knitted structures, based on the desired end use.
- Determining the physical characteristics of both fibers and fabrics, in particular their mechanical characteristics, molecular orientation, porosity, and specific surface area, as well as their functional characteristics, depending on their planned end uses.
- Developing, optimizing, and designing the respective fabric components of technical textile segments and fabricating prototypes from these invented fabric components.
- Examining and assessing the effectiveness of manufactured fabric in responding to functional behaviors, such as protective or hydrophilic properties.
- Transferring all of these scopes from laboratory-scale research to an industrial level for product development.

REFERENCES

1. Dasaradhan B, Das BR, Sinha MK, Kumar K, Kishore B, Prasad NE. 2018. A brief review of technology and materials for aerostat application. Asian Journal of Textile 8(1):1–12. https://doi.org/10.3923/ajt.2018.1.12
2. Seiko J, Pandit P, Pandey R. 2019. Chickpeahusk—A potential agro waste for coloration and functional finishing of textiles. Industrial Crops & Products 142:111833. https://doi.org/10.1016/j.indcrop.2019.111833
3. Pandit P, Jose S, Pandey R. 2020. Groundnut Testa: A potential agro residue for single bath dyeing and protective finishing of cotton fabric. Waste and Biomass Valorization. https://doi.org/10.1007/s12649-020-01214-y
4. Joshi S, Kambo N, Dubey S, Shukla P, Pandey R. 2021. Effect of onion (Allium cepa L.) peel extract-based nanoemulsion on anti-microbial and UPF properties of cotton and cotton blended fabrics. Journal of Natural Fibers. https://doi.org/10.1080/15440478.2021.1964127
5. Badanayak P, Jose S, Dubey R, Pandey R. 2022. Tools and methods for handling and storage of museum textiles. Handbook of Museum Textiles 1:161–180. https://doi.org/10.1002/9781119983903.ch8
6. Bhagat S, Sachdeva K. 2022. Ideal storage conditions for museum textiles. Handbook of Museum Textiles 1:143–160. https://doi.org/10.1002/9781119983903.ch7
7. Ahmad S, Ahmad F, Afzal A, Rasheed A, Mohsin M, Ahmad N. 2015. Effect of weave structure on thermo-physiological properties of cotton fabrics. AUTEX Research Journal 15(1):30–34. https://doi.org/10.2478/aut-2014–0011D
8. Redmore N. 2014. Open to change: Embracing nature and the fragility of design. An investigation into outdoor seating materials through the practice of leno weaving (Doctoral dissertation, University of Huddersfield).
9. Saha J, Rahman M, Kabir MR, Emtiaz ANM, Islam MR, Jafor A. 2017. Study on manufacturing process of leno weave by modification of hand loom. Journal of Science and Technology 7:157–170.

10. Shaker K, Nawab Y, Ayub Asghar M, Nasreen A, Jabbar M. 2020. Tailoring the properties of leno woven fabrics by varying the structure. Mechanics of Advanced Materials and Structures 27(22):1865–1872. https://doi.org/10.1080/15376494.2018.1527964

11. Jaafaripour M, Amid E, Iranmanesh F. 2017. Designing museum of anthropology with attitude of preserving regional indexes in Hormoz Island. Journal of History Culture and Art Research 6(3):1392–1406. https://doi.org/10.7596/taksad.v6i3.1010

12. Leader M. 2001. Conservation of a crewelwork bed curtain. Conservation Journal 39.

13. Adak B, Joshi M. 2018. Coated or Laminated Textiles for Aerostat and Stratospheric Airship. Wiley-Scrivener Publishing, Beverly, MA, 257–287.

14. Brzeziński S, Rybicki T, Karbownik I, Śledzińska K, Krawczyńska I. 2010. Usability of a modified method for testing emissivity to assess the real shielding properties of textiles. Fibres & Textiles in Eastern Europe 18(5):76–80.

15. Kawabata S, Inoue M, Niwa M. 1992. Non-linear theory of the biaxial deformation of a triaxial-weave fabric. Journal of the Textile Institute 83(1):104–119.

16. Pandey R, Jahan S, Goel A. 1999. Hand knitted fabric of merino wool. Indian Textile Journal 110(2):122–125.

17. Pandey R, Prasad GK, Dubey A, Arputhraj A, Raja A SM, Sinha MK, Jose S. 2022. Tellicherry bark microfiber: Characterization and processing. Journal of Natural Fibers 19(16):13288–13299. http://dx.doi.org/10.1080/15440478.2022.2089432

18. Chapman R. 2010. Applications of Nonwovens in Technical Textiles. Elsevier, Cambridge, UK.

19. Sinha MK, Das BR, Srivastava A, Saxena AK. 2014. Study of electrospun chitosan nanofibrous coated webs. Journal of Nano Research 27:129–141. https://doi.org/10.4028/www.scientific.net/JNanoR.27.129

20. Gong X, Chen X, Zhou Y. 2018. Advanced weaving technologies for high-performance fabrics. In High-Performance Apparel. Woodhead Publishing, 75–112. https://doi.org/10.1016/B978-0-08-100904-8.00004-3

21. Potluri P, Nawaz S. 2011. Developments in braided fabrics. In Specialist Yarn and Fabric Structures. Woodhead Publishing, Cambridge, 333–353.

22. Wambua PA, Anandjiwala R. 2011. A review of preforms for the composites industry. Journal of Industrial Textiles 40(4):310–333. https://doi.org/10.1177/1528083709092014

23. Bilisik K, Karaduman NS, Bilisik NE. 2016. Fiber architectures for composite applications. In Fibrous and Textile Materials for Composite Applications. Springer, Singapore. https://doi.org/10.1007/978-981-10-0234-2_3

24. Farooq AS, Zhang P. 2021. Fundamentals, materials and strategies for personal thermal management by next-generation textiles. Composites Part A: Applied Science and Manufacturing 142:106249. https://doi.org/10.1016/j.compositesa.2020.106249

25. Morris H, Murray R. 2020. Medical textiles. Textile Progress 52(1–2):1–127. https://doi.org/10.1080/00405167.2020.1824468

26. Kang TJ. 2007. Smart textiles in modern life. In Proceedings of the Korean Fiber Society Conference, The Korean Fiber Society, 3–6.

27. Ahmed A, Hossain MM, Adak B, Mukhopadhyay S. 2020. Recent advances in 2D MXene integrated smart-textile interfaces for multifunctional applications. Chemistry of Materials 32(24):10296–10320. https://doi.org/10.1021/acs.chemmater.0c03392

28. Brewer SA, Willis CR. 2008. Structure and oil repellency: Textiles with liquid repellency to hexane. Applied Surface Science 254(20):6450–6454. https://doi.org/10.1016/j.apsusc.2008.04.053

29. Gargoubi S, Baffoun A, Harzallah OA, Hamdi M, Boudokhane C. 2020. Water repellent treatment for cotton fabrics with long-chain fluoropolymer and its short-chain eco-friendly alternative. The Journal of the Textile Institute 111(6):835–845. https://doi.org/10.1080/00405000.2019.1664796

30. Shi X, He J, Lu X. 2020. Functional structural color dye with excellent UV protection property. Dyes and Pigments 175:108142. https://doi.org/10.1016/j.dyepig.2019.108142
31. Shi H, Xin JH. 2007. Cosmetic textiles: Concepts, application and prospects. 2007-09-05T05:32:52Z
32. Cheng SY, Yuen CWM, Kan CW, Cheuk KKL. 2008. Development of cosmetic textiles using microencapsulation technology. Research Journal of Textile and Apparel 12(4):41–51. https://doi.org/10.1108/RJTA-12-04-2008-B005
33. Sinha MK, Das BR. 2018. Chitosan nanofibrous materials for chemical and biological protection. Journal of Textiles and Fibrous Materials 1:1–13. https://doi.org/10.1177/2515221118788370
34. Thilagavathi G, Bala SK, Kannaian T. 2007. Microencapsulation of herbal extracts for microbial resistance in healthcare textiles. Indian Journal of Fibre & Textile Research 32:351–354. http://nopr.niscpr.res.in/handle/123456789/421
35. Pieniak D, Walczak A. 2019. Preliminary studies on scratch resistance of the face shields surface of firefighting helmets. Autobusy–Technika, Eksploatacja, Systemy Transportowe 20(1–2):322–326. https://doi.org/10.24136/atest.2019.059
36. Mikhaylova A, Liesenfeld B, Moore D, Toreki W, Vella J, Batich C, Schultz G. 2011. Preclinical evaluation of antimicrobial efficacy and biocompatibility of a novel bacterial barrier dressing. Wounds 23(2):24–31.
37. Zamora-Mendoza L, Guamba E, Miño K, Romero MP, Levoyer A, Alvarez-Barreto JF, Machado A, Alexis F. 2022. Antimicrobial properties of plant fibers. Molecules 27(22):7999. https://doi.org/10.3390/molecules27227999
38. Kosior E, Braganca RM, Fowler P. 2006. Lightweight compostable packaging: Literature review. The Waste & Resources Action Programme 26:1–48.
39. Konwarh R, Karak N, Misra M. 2013. Electrospun cellulose acetate nanofibers: The present status and gamut of biotechnological applications. Biotechnology Advances 31(4):421–437. https://doi.org/10.1016/j.biotechadv.2013.01.002
40. Pandey R, Pandit P, Pandey S, Mishra S. 2020. Solutions for sustainable fashion and textile industry. In Pintu Pandit et al. (eds) Recycling from Waste in Fashion and Textiles: A Sustainable & Circular Economic Approach. Scrivener Publishing, Beverly, MA, USA, 33–72.
41. Esthappan SK, Sinha MK, Katiyar P, Srivastav A, Joseph R. 2013. Polypropylene/zinc oxide nanocomposite fibers: Morphology and thermal analysis. Journal of Polymer Materials 30(1):79–89.
42. Liu Y, Kumar S. 2014. Polymer/carbon nanotube nano composite fibers—A review. ACS Applied Materials & Interfaces 6(9):6069–6087. https://doi.org/10.1021/am405136s
43. Sharma A, Mandal T, Goswami S. 2021. Fabrication of cellulose acetate nanocomposite films with lignocelluosic nanofiber filler for superior effect on thermal, mechanical and optical properties. Nano-Structures & Nano-Objects 25:100642. https://doi.org/10.1016/j.nanoso.2020.100642
44. Narani SS, Abbaspour M, Hosseini SMM, Aflaki E, Nejad FM. 2020. Sustainable reuse of Waste Tire Textile Fibers (WTTFs) as reinforcement materials for expansive soils: With a special focus on landfill liners/covers. Journal of Cleaner Production 247:119151. https://doi.org/10.1016/j.jclepro.2019.119151
45. Formela K. 2021. Sustainable development of waste tires recycling technologies–recent advances, challenges and future trends. Advanced Industrial and Engineering Polymer Research 4(3):209–222. https://doi.org/10.1016/j.aiepr.2021.06.004
46. Fazli A, Rodrigue D. 2022. Sustainable reuse of waste tire textile fibers (WTTF) as reinforcements. Polymers 14(19):3933. https://doi.org/10.3390/polym14193933

47. Kang J, Zhang B, Li G. 2012. The abrasion-resistance investigation of rubberized concrete. Journal of Wuhan University of Technology-Materials Science Edition 27(6):1144–1148. https://doi.org/10.1007/s11595-012-0619-8

48. Svoboda J, Dvorský T, Václavík V, Charvát J, Máčalová K, Heviánková S, Janurová E. 2021. Sound-absorbing and thermal-insulating properties of cement composite based on recycled rubber from waste tires. Applied Sciences 11(6):2725. https://doi.org/10.3390/app11062725

49. Thai QB, Le DK, Do NH, Le PK, Phan-Thien N, Wee CY, Duong HM. 2020. Advanced aerogels from waste tire fibers for oil spill-cleaning applications. Journal of Environmental Chemical Engineering 8(4):104016. https://doi.org/10.1016/j.jece.2020.104016

4 Fibers from Agro-Residues for High-Value Technical Textiles

4.1 INTRODUCTION

The importance of natural fibers can be understood from the fact that the development of technical fibers with high performance focuses on biomimicking cotton, wool, and silk [1]. According to the Japanese Chemical Fibers Manufacturers Association (JCFA), agricultural fibers are an important asset for technical textiles [2]. Agro-waste includes farm waste such as discarded parts of crops, field drains, fertilizer runoff, and poultry and dairy farm waste [3]. Fibers from agricultural farms have been used as technical textiles for centuries in the form of ropes, marine products, packaging, furnishing, and geotextiles. Cotton, jute, hemp, flax, and sisal fibers are the earliest known fibers used in technical textile segments to prepare various products such as coarse canvas, carpet, linoleum, furniture, basket, rope, and twine [4]. The earliest uses of technical textiles are tents and tarpaulin with water-resistant finishes. Goatskin leather water bags known as *mashak*s were used until the early 20th century during warfare. Silk was used as surgical sutures and sewing thread in technical textile components. Wool fiber with good insulation properties finds its application in protective clothing at higher altitudes. The flame-retardant characteristics of wool make it an ideal choice for fabricating furnishings in the inflammable chemical manufacturing industry. Fibers used in technical textiles are made with special treatments or finishes applied or a combination of both to make them high-performance fabrics. Technical textile performance and characteristics are superior to those of fabrics being used for just apparel and decorative purposes [4].

Farm fields fill the ever-increasing needs for food, textiles, building construction, paper pulp, and energy. Rapid industrialization has also contributed to further exploitation of plant resources. The practice generates an increasing amount of plant by-products and residues discarded in the fields and industry, which are the source of raw materials for bioenergy, dyeing, finishing, handicrafts, soil additives, and biofertilizers [5]. Standard agricultural practices ensure timely harvest of plants and felling of trees to utilize plant parts such as seeds, grains, fruits, and flowers. A number of plant parts and post-harvest remains of plants are currently underutilized but have significant use in medicines, bioenergy, textiles, paper, colorant for food, furniture, walls, and fibers. Farm residues are biodegradable, in contrast to synthetic fiber manufacturing and products that disseminate toxic substances, polluting the soil, water, and environment. Sustainability issues have been a focus on many platforms, including ASEAN vision 2025 and the

UN 2030 Agenda, for ecological and economic prosperity through the promotion of sustainable practices [6, 7]. It is time to initiate all-out efforts to utilize fibers from farm waste that are nonpolluting and user friendly compared to synthetic fibers. Widespread promotion of farm waste utilization for high-value fibers will enhance the income of farmers and fill the demands of environmentally concerned consumers. An Italian consumer survey from Cameron and Huppert shows consumers are willing to pay significant premium prices, from 64 to 128% more, on bio-textile products. The result will boost bio-textile products in the European economy and reduce dependency on fossil fuels [8]. Studies on *Boops boops* and nori have confirmed the consumption of anthropogenic fibers and microplastics by humankind through unprocessed, factory-processed, and commercially packaged fish species [9–12]. Furthermore, 33% of the fibers present in indoor environments are predominantly microplastics containing polypropylene and petroleum products [13]. Microplastics floating in the environment pose a serious threat to health [14]. Inhaled microplastics carry pollutants with them and have adverse effects on human health [15].

Agricultural crop residues contain 30–60% cellulose, 8–25% hemicellulose [16] and 12–28% lignin [17]. Higher cellulose is preferred in textile, paper, and various fibrous applications. Lignin is a tough component that decomposes at high temperature above 900 °C. Lignin content influences fiber structure, stiffness, and rate of hydrolysis. A hollow lumen at the core of plant fiber is filled with air, contributing to acoustic absorption and thermal insulation. The lumen structure is associated with water uptake of the fiber. The bigger the lumen, the higher the moisture content of the fiber [18]. A large human-made cellulosic fiber producing industry require about 1600 kiloton of dry agricultural biomass. The required biomass should have a minimum of 30% fiber yield [3]. Industry essentially requires more than one source of plant biomass to meet this demand. Plant fibers have the potential to cater to the large-scale demand of the technical textile industry due to their abundant availability and renewability [19–23].

The chapter emphasizes ways to utilize agro-wastes such as oil flax, pina, coir, typha, milkweed, and lotus for various technical textile segments [24–26]. Plant fibers are not only a potential substitute for existing synthetic fibers but also have high economic and environmental impacts. A description of all such aspects, including fiber extraction and utilization, is also presented.

4.2 CLASSIFICATION OF AGRO-WASTE

4.2.1 PRIMARY RESIDUES (FIELD RESIDUES)

Primary residues are those discarded after harvesting crop/fruit/food grains/feathers [27–31]. Examples are leaves, stems, straw, prunings, seed pods, roots, and inflorescence, which are the source of fibers and color (Table 4.1). The approximate agro-residue availability in India is 600 MT/year. Prevalent agro-residue management practices and their impact on the environment are presented in Table 4.2 [3].

TABLE 4.1
Agro-Residue Classification

Agro-Residue Classification		
Primary residue (farm residue)	Farm residue and dumps	Plant stalks, leaves, flowers, luffa, inflorescence, seed husk
	Farm drains and dumps	Organic manure, fertilizer, pesticide run-off
Secondary residue (mill residue)	Food and agro-industrial processing residue	Seed husks, hulls, bagasse, molasses, fruit pulp, wood peelings, shive
	Textile industry by-products	Fiber lint, jute caddies

TABLE 4.2
Current Practices Followed to Manage Agro-Residues

Farm Residue		
Management	Current Uses	Impact
Crop residue	Fodder, organic manure, fuel, animal bedding, mulching, bio-based fibers	Pollution, breathlessness, eye problems
Cow dung and poultry waste	Biogas, cow dung cake as fuel, organic manure	Saves fossil fuel and the environment
Crop residue burning	-	Pollution, asthma, breathlessness, eye problems
Fertilizer and pesticide run-off	-	Excess chemicals increase soil toxicity, making it unfit for crop production

4.2.2 SECONDARY (MILL RESIDUE)

4.2.2.1 Agricultural Industrial Processing Residue

This refers to plant parts left after obtaining fruit, vegetables, and food grains. Examples are seed testa of nuts and pulses, rice bran, fruit rinds, bagasse, molasses, luffa, empty fruit brunch (palm oil mill residue), and the fibrous parts of fruits and vegetables. Secondary residue is highly valued in the field of technical textiles. India generates around 350 MT/year of agro-industrial residue. Industrial processing of primary crop residues and bast fibers generates wood scraps, shives, and large amounts of fiber lint. Wood scraps and shives are used in composites, paper making, and energy sectors [3].

4.2.2.2 Textile Industry Processing Residue

The textile industries generate carding and spinning waste called pre-consumer waste, lint, or shoddy fiber [32–34]. Fiber lint is utilized as a raw material for humanmade fibers, carpets, briquets, acoustics, fillers, sanitary napkins, and soil enhancers. Thailand and Indonesia provide 75 to 150 kilotons of cellulose pulp per year to small and medium cellulose processing plants [3].

4.3 AGRO-RESIDUE UTILIZATION

Agro-residue is generally utilized by farmers as fuel, animal feed, roof thatching, and a utensil scrubber and mixed with cow dung cake for fuel and energy requirements. Innovative uses of agro-residue [35–40] are presented in Table 4.3. Unutilized large-scale agro-wastes contribute to solid waste generation and are discarded along with other household wastes. Agro-waste burning in fields to get rid of it for sowing of the next crop is the most common practice among farmers. Crop residue burning is a cause of air pollution and hazy atmosphere, especially in the winter post-paddy harvesting [3].

TABLE 4.3

Agro-Residue Fibers and Their Innovative Uses in Technical Textiles

Natural Fibers from Agro-Residue	Product Fabrication/ Use	Technical Properties	Reference
Bacterial cellulose	Clothtech	Contact angle 28.53° hydrophilic, Water holding capacity (320%)	39
Bagasse, sisal fiber maize husk, cotton rags, wastepaper	Handmade paper, biofuel production	High tensile strength	30
Banana fiber	Paper, cellulose macromolecules	Composite	21
Coir	Mulch mats, composite	Hydrophilic	26
Calotropis gigantea	Reinforced composite	50% bacterial reduction	25
Chicken feathers	Cement mixing	Cement setting, hydration, prevention of crystallization	Mendoza, Grande, & Acda 2019
Cornstalk fiber	Biofibers	Good tensile strength	17, 19
Cotton linters, bagasse, cardboard	Wet-spun regenerated cellulose fibers	Nonwoven rolls	28
Corn, wheat, rice, sorghum, barley, sugarcane, pineapple, banana, and coconut	Wet-spun fiber	High tensile strength	28, 32
Dracaena steudneri Egler (Serte) leaf fiber	Commercial microcrystalline cellulose	73.9% crystallinity, 30–60 micron	27
Eucalyptus wood production waste	Dye for cotton, wool, nylon	Good fastness properties	22

Natural Fibers from Agro-Residue	Product Fabrication/ Use	Technical Properties	Reference
Food packaging waste, tetra pack	Composite panel	Thermal insulation	35
Keratin-alginate crosslinked	Keratin from chicken feathers	Tough fibers for tissue engineering	29
Keratin fiber, chicken feathers	Cement hardening (tenacity: 0.5 cN/tex, 3.5 MPa)	6% fiber improved initial (46%) and final (54%) setting time	31
Mulberry wood waste	Antioxidant activity (83.5%) and colorfastness	Good fastness properties	34
Pineapple	Composite panel	Bending property	24
Rice straw fiber, rice husk powder, rape straw, rye straw, hemp straw, carrot leaves, wheat straw	Paper, bioenergy, composting, composite, thatching, biofibers	Strength, increasing the adhesive strength between the fiber and matrix	32, 36
Roselle	Reinforced composite	High tensile strength	18, 23
Rice straw fiber	Building construction	Strength	23
Waste cotton textiles	Dashboard panel, automotive	Low water diffusion coefficient than wood	37
Wheat straw	Feed stock, energy, structural composite	Excellent strength and stiffness	20, 32
Wool waste, chicken feathers	Biocomposites, automotive, building interiors, mulch mats, building insulation, reinforced polymer, oil spill sorbents and soil fertilizers, carbon nanofibers, scaffolds for tissue engineering, cement mix	Activated carbon, bio-based thermoplastics, biomedicine, biofertilizers, thermally stable, setting and hydration characteristics	31, 33, 38.

4.3.1 AGRO-PRODUCTS

Agriculture and textiles are inseparable because agriculture is the basic source of raw materials for textiles, benefitting the industry and stakeholders. Agricultural fibers have been used in textiles since ancient times. Agro-textiles cater to the needs of agriculture, horticulture, dairy, and fish farming. Agro-textiles are used for the protection of plants, ensuring higher productivity of food grains, vegetables, and fruits. Agro-textiles activate fruit and grain protection, growth, collection, weed suppression, and light reflection. The global market share of agro-textiles was 6% (8.4 USD) of the total technical textile segment in FY 2021. Synthetic fibers like polyethylene, polypropylene, and polystyrene are extensively used in agro-textiles. Synthetic fibers and films are a threat to the environment and human health. The search for alternatives led stakeholders to lignocellulosic fibers [6, 17]. Cellulosic fibers are abundantly available in the form of agro-residue [41–44]. Common practice for field mulch application is spreading plastic sheets on fields, which pose disposal problems. Recent technology utilizing cellulosic fiber, wheat bran, and seaweed nanofiber film (biodegradable) sprayed on crop fields ensures crop protection, air and water permeability, and end-of-life management dynamics [6]. Seedling transportation from the nursery is generally done in polyethylene bags. The bags are slashed and thrown away while transplanting seedlings in fields or pots. Cellulosic fiber-based seedling bags or root ball sleeves containing plants are directly placed in fields or pots. Fiber sleeves and root balls provide nourishment and air circulation to roots, ensuring healthy growth. The practice is low cost, less strenuous, and pollution free [45]. Mulch mats made of coir protect seeds and seedlings from rain and wind, facilitating growth. Mats also prevent soil erosion and nutrient leaching, maintaining soil fertility. Nonwoven mulches of flax, hemp, jute, wool, and cotton are prepared by the needle-punched method. Application of fertilizer-coated agro-fibers on crop fields improves crop health and productivity [46, 47]. Furthermore, agro-textiles turn into biomass that degrades, enriching soil after serving for a span of time. In summary, agro-textiles help in crop protection, crop growth, soil enrichment, fruit and grain collection, packaging, and transportation (Table 4.4). Turboelectric nanogenerator (TENG) yarn embedded in an agro-textile net is capable of energy generation from rain and helps crop growth and productivity. A product range for agro-textiles, their usefulness, and suitable fibers are presented in Table 4.4 [17].

TABLE 4.4
Textile Fibers in Agrotextile Applications [41–47]

Agrotextile	Applications	Advantages
Coir	Bed blankets on seedling trays, biorolls, basket linters, bed blankets, roof green mats, grow sticks, erosion control blankets, coco logs, supporting rubberized poles for climbers	Seed germination, wetland restoration, coir nonwoven roll covered with coir pith promotes root growth
Coir coated with kerosene	Nonwoven rolls, mulches	Prevents early decaying of fibers merged with soil. Improves strength and elongation of the coated fibers
Cotton	Nonwoven agrotextile sheets, cotton root balls, harvesting nets, sampling bags, grain packaging	Improved radish seed germination rate in less time
Jute agrotextile	Baler twines to wrap crops, seedling sleeves, grain packaging	Improved broccoli yield, volume, and weight

Agrotextile	Applications	Advantages
Nanocellulose	Mulching nanofilm made of cellulose and animal protein sprayed on cultivated fields	Protects crop and later biodegrades in fields, mitigating the pollution caused by discarding plastic mulch
Plant fibers (flax, hemp, sisal, banana, pineapple)	Nonwoven agrotextile sheets, sampling bags, baler twines, grain packaging, windbreakers, crop covers, fruit covers, anti-hail nets, fishing nets, shade nets, anti-insect nets, bird protection nets, vermin beds	Degrades over time and fertilizes soil, biodegradable, renewable, neutral CO_2 emission, high mechanical strength, hydrophilic (absorbs and releases moisture to the crop per plant requirements), preserves soil moisture, protection during transportation, maintains soil on root surface, low density, elongation, elasticity, low cost
Ramie fiber	Nonwoven ramie film is padded at the bottom of rice seedling trays	Improved rice seedling growth, nitrogen
Rice straw + Fertilizers (NPK)	Soil enhancers, mulching	Glycoprotein secretion in soil improved up to 5.67%
Textile covers	Mulch film	Retains moisture in the range of 15–60 to 100–500 g/m^2
Wool waste	Mulch mats	30% grain yield of tomato and pepper, improves grain yield, barley and fodder growth, turns into fertilizer

4.3.2 Biocomposites

A composite is defined as the combination of two or more solid phases with several desirable properties, including strength, stiffness, and resistance to heat. Concern for the environment has attracted researchers and industrialists towards plant fiber and green epoxy–based biocomposites with renewability, biodegradability, and low weight [48–52]. An advantageous admixture of solid phases to bring essential features suitable for end use requirements is key to fiber-based biocomposite fabrication. Cellulosic fibers composed of cellulose, hemicellulose, lignin, pectin, and waxes are themselves a perfect example of composite structure. Cellulose contributes to the tensile strength and stability of plant fibers. Hemicelluloses with lesser orientation (low crystallinity and chains not perfectly aligned) and lower degree of polymerization than cellulose aid in binding microfibrils. Lignin and pectin provide strength, binding, stiffness, and high thermal stability to the fiber [16, 18, 53]. Natural fibers, hydrophilic in nature, have less compatibility with hydrophobic polymer matrices. Fiber surface modification to suit industrial needs is done by eco-friendly plasma and enzymatic treatments [54]. Interfacial strength and bonding between fibers and the polymer matrix are improved by alkalization of fiber prior to composite manufacturing. Composite fabrication methods are presented in Table 4.5 [53]. Traditionally, high-strength composite fabrication is done by using glass, carbon, and aramid fiber reinforced with epoxy, plastics, and polyester. The fiber–matrix interface of a composite is classified further on the basis of the fiber matrix as a metal–matrix, ceramics–matrix, or synthetic polymer–matrix composite. Synthetic fiber composites produce plastic waste, which is

TABLE 4.5
Fiber-Reinforced Composite Fabrication Techniques

Composite Fabrication	Technique	Articles	Advantages
Open mold	*Hand lay up*—liquid resin is poured on fiber bed and spread by hand, roller, brushes, then cured, and several layers are added *Spray up*—liquid resin is sprayed with a spray gun, then cured and cooled	Large and less complex shapes: boat hulls, bathtubs, shower and spa units	Lower labor cost
Resin or liquid molding	*Resin transfer molding*—Low-viscosity resin is pumped into mold + preformed fiber reinforcement and mold is closed	Thick composite fabrication	Simple method
	Reaction injection molding—resin and catalyst are pumped under pressure into a mold containing fiber preformed layer	Structural part fabrication	Low cost, faster automated production
	Vacuum assisted resin transfer molding—resin is drawn on fiber preform under vacuum conditions	Large and complex part fabrication	Low cost, fast
	Resin film infusion—High-viscosity resin is drawn on fiber lay up under heat, pressure, and vacuum	Boat and aerospace laminates	Uniform resin distribution, low voids, reproducibility
Compression molding	Resin placed on mold + fibers are pressed using hydraulic pressure	Composites in various sizes, thicknesses, complex parts, volumes	Faster production with high strength and quality
Injection molding (1960)	Nylon and glass fiber are injected into molds at low pressure	Parts of machines such as screws	Fast method

Composite Fabrication	Technique	Articles	Advantages
Filament winding	Fibers are run through resin bath in long cylindrical mandrel	Aerospace composite material	Continuous fabrication in predetermined configuration, low material cost, repeatable
Pultrusion method	Fibers are run through heated and wet resin bath and shaped dies to form long pieces	Ladders, molding	Continuous linear fabrication, highly automated

non-biodegradable and harmful for soil and human health. Ecological issues and petroleum shortages are increasing the need for use of natural textile fibers for biocomposite manufacturing. The advantages of natural fibers are its non-toxicity, biodegradability, low density, cost effectiveness, and recyclability [19]. Natural fibers are also recommended for their renewability, recyclability, flexibility, and good sound insulation. Increasing awareness of the benefits of natural fibers in textiles is encouraging their use in composite manufacturing [55, 56]. The advantages are non-toxicity and biodegradability. The earliest examples of the use of agro-residue for composite structures dates back to 1500 BC in the form of bricks, pottery, and boats made with combinations of plant straw and mud. Reinforcement with straw provided strength and flexibility to building structures [53]. Fiber-filled polymer composites are lightweight due to low fiber density. Common plant fibers used for biocomposite fabrication are flax, jute, hemp, banana, luffa, betel nut, ramie, sisal, and sugarcane [57]. Agro-residue fibers such as coir, pineapple, jute stick, and palm are potential sources of reinforcing composite in manufacturing lightweight boats [16], car panels, and dashboards and in aerospace, marine, electrical, sports, and household appliance applications [58]. The uses of biocomposites made with natural plant fibers range from building technology to acoustics, automobiles, and electrical goods.

4.3.2.1 Building Construction

Building construction, agriculture, and textiles are the three most important economic activities in India. Among the three, construction activities are the highest contributor of CO_2 emissions (39%). To achieve carbon neutrality in the building and construction sector, agro and forest waste in cement mixtures has attracted the attention of researchers, policy makers, and builders [59]. Agro-residue and agro-based textile fibers are advantageous in construction activities to make products lightweight with improved compressional strength. Conventional building construction has been dependent on cement, metals, bricks, and wood generated by the earth but involves energy requirements and causes environmental pollution. Natural fibers are lightweight and provide strength and stiffness to the composite, whereas fiberglass is brittle [60]. Sustainable construction using agro-residue fibers save precious natural resources and energy. A rice husk ash, lime, and cement mixture is an alternative low-cost housing solution in rural areas. Bricks made with a mixture of clay, sand, and rice husk ash increase the compressive strength compared to conventional bricks. A brick mixture containing 15% rice husk sintered at 1100 °C was low in density and high in strength. Agro-residue utilized for concrete mixing improves thermal stability [61]. Agro-residue and its by-products improve strength, as well as reducing CO_2 emissions, water penetration, and density of the concrete.

Green buildings made with the incorporation of agro-residue fibers in concrete structures and brick manufacturing save energy and have relatively light weight, good insulation properties, and high compressive strength. Thermal insulation through construction material protects the building from sun heat and thus promotes a comfortable temperature. A cool interior of a building will minimize the duration of running an air conditioner. Fiberglass dominated as an insulation material in construction for centuries. However, the harmful effects of fiberglass handling and use on human skin, lungs, and environment prompted stakeholders to adopt and evolve alternative sustainable building materials. Agro-residue fibers are renewable, biodegradable, and abundantly available [56]. The thermal conductivity of insulation boards made of agro-residue fibers is comparable to fiberglass (Table 4.6). The thermal conductivity of insulation boards below

TABLE 4.6
Agro-Residue Applications for Buildtech

Fibers	Substitute for	Binder/Density (kg/m³)	Thermal Conductivity in Watts per Meter Kelvin (W/m-K)	Reference
Banana bunch	Fiber boards	Natural binder/1000	-	17
Coffee husk + hull	Panel boards		0.110	17
Coir + durian peel	Particle boards	311–611	0.0728–0.1117	48
Corn husk	Particle boards water absorption = 11–14%	310	-	49
Cotton stalk fiber	Particle boards	Binderless/150–450	0.0585–0.0815	50, 55
Kenaf fiber	Particle boards High-strength low-density, saves energy, water absorption 13%	Binderless/150–200	0.040–0.065	51, 56

Fibers	Substitute for	Binder/Density (kg/m³⁾	Thermal Conductivity in Watts per Meter Kelvin (W/m-K)	Reference
Pineapple leaf fiber	Natural rubber latex	210	0.035	52
Rice husk ash	Particle boards, insulating material, High mechanical strength, low thermal conductivity	Urea formaldehyde/0.000229	-	49
Typha fiber	Insulation boards	Methylene diphenyl diisocyanate (MDI)/200–400	0.0438–0.0606	56
Wheat straw	Particle boards, insulating material	150–250	0.0481–0.0521	55
Wool waste	Insulation boards	-	0.044–0.057	33

0.25 W/mK is considered suitable for building construction applications. The lower the density of the boards, the greater the insulation property due to low thermal conductivity. Low-density insulation boards have greater voids in them compared to high-density boards of the same thickness. Large voids of the boards are filled with air, which hinders thermal conduction [52].

4.3.2.2 Acoustics

Noise is the third largest source of pollution affecting our environment, economy, and health [57]. The majority of mechanical and electrical machineries used in industries, vehicular movement, and households cause vibrations and noise pollution. Various sound absorbers and soundproof panels have been developed to minimize the problem of unwanted vibrations and noise. Sound-absorbing panels are hard boards made of materials that reduce the reflection and transmission of vibrations and sound. The acoustic behavior of the material is influenced by its thickness; surface hindrance capabilities; tortuosity; and characteristics of the reinforced fibers such as size, density, porosity, and air resistivity [62]. Materials used are of two types: porous, which is further divided into fibrous (natural and anthropogenic fibers), granular (wood chips), and cellular (foams). The second type is microperforated and a combination of the two types. Sound absorption and air flow resistivity are improved in composites reinforced with a decrease in fiber diameter [62–64]. Panel boards densely covered with plant fibers amplify sound absorption because increasing the number of fibers presents a tortuous path for sound energy, causing higher friction and energy loss. Furthermore, in the case of plant fibers, the more fibers, the greater the presence of lumen, which helps in breaking, scattering, and removing sound energy. The presence of a higher fiber content with high tenacity results in higher impact strength (breaking resistance) and flexural strength (resisting deformation) of fiber-reinforced composites. Higher plant fiber content in fiber-reinforced panels also increases thermal diffusivity (heat spread through material) and thermal resistance (prevention of heat loss) but decreases thermal conductivity (fiber-intrinsic property to transfer heat) and thermal expansion (dimensional changes) [57]. The sound absorption coefficient is increased with higher porosity, surface impedance, and air cavities in bonded panel boards [62]. The advantages of airflow resistivity, besides acoustical importance, are in filtration and comfort properties [64]. Synthetic fibers are commonly used as reinforced material in particle boards for acoustic solutions. However, due to renewability and environmental friendliness, natural fibers have now taken center stage as a prominent sound-absorbent material [65–72] (Table 4.7). Flax, jute, and ramie fiber composites demonstrated higher sound-absorbing capacity than glass and carbon fiber-reinforced composites at a frequency range of 1000 Hz [57, 73]. Problems associated with natural fibers in acoustical composites are their hydrophilicity and irregular shape (thick and thin places) [65]. Certain treatments on fiber such as plasma, ultrasonic, and grafting improve the fiber and resin compatibility. High sound-absorbing

TABLE 4.7
Sound Absorbent Coefficient of Plant Fiber–Reinforced Composites

Composite Material	Frequency Range (Hz)	Sound Absorbent Coefficient	Reference
Banana + epoxy	500–6000	0.11	[72]
Ramie + polylactic acid	250–1600	0.089–0.353	[74]
Ramie + polylactic acid	500–6000	0.121	[74]
Wood and plant husk ash + formaldehyde	400–3200	0.8–0.9	[66]
Flax epoxy	2000	0.6–0.65	[67]
Banana + polyester	4000	0.685	[69]
Bagasse + polyester	4000	0.6338	[69]
Kenaf + urea formaldehyde	2000	0.065	[68]
Rice straw + polypropylene	2000	0.008	[68]
Sisal + polylactic acid	2000	0.085	[70]
Coir + epoxy	6000	0.78	[70]
Coir + epoxy	0–1600	0.09–0.183	[57]
Cotton + epoxy	0–1600	0.059–0.125	[57]
Sugarcane + epoxy	0–1600	0.081–163	[57]
Bagasse + epoxy	6000	0.75	[71]
Kenaf + epoxy	6000	0.78	[71]
Flax + epoxy	250–10000	0.96	[73]
Basala wood + epoxy	250–10000	0.58	[73]

capability is reported in biocomposites made with reinforced coir, kenaf, rice straw, and teal leaf fibers [57].

4.3.2.3   Automobiles

Composites are superior in quality compared to single materials. Composites are required in aircraft fabrication in tons of quantities. Bast fibers such as flax and hemp with high strength and low density are suitable for reinforcement in aircraft fuselage [1]. The automobile industry is a huge buyer of technical textile fabrics. Smart interiors, dashboard covers, doormats, headliners, and upholstery are selected to provide warmth and comfort. The use of technical textiles in automobiles is functional, as they are low in density, flexible, and foldable yet cozy. Currently 35 kg of textile fibers (approximately 45 sq m) is used in a car, mainly in carpets and mats, followed by upholstery [75]. Durability, comfort, wrinkle recovery, shape retention, appearance, antistatic, easy care, pill resistance, air permeability, acoustic insulation, smoothness, and flame-resistant properties are the requirements for car interior textile material.

4.3.3   BIOSORPTION

Agro-residue acts as a phytoremediator and thus detoxifies soil. Heavy metal pollutants are absorbed by plant parts. Flax and typha were found to be the most effective bio-absorbents. They absorb heavy metals from polluted soil [76, 77]. Water pollution caused by discharge of untreated effluents harms the ecological balance of aquatic bodies. Industrial effluents containing dyes, finishing chemicals, and microfibers from household washing machines cause heavy metal contamination, which is detrimental to human and aquatic organism health. Similarly, extensive use of pesticides, fertilizer runoff discharge, and groundwater extraction for improved production pollutes fields and drains with chemicals. The use of fertilizers in crop fields contributes to the carbon footprint of cultivated products [3, 6]. Microplastics containing dyes and finishing additives make their way to our dining tables [9–12]. Fibers are macromolecules with active sites to attract functional groups present in color, but the use of synthetic fibers for color removal poses an environmental threat. Natural fibers have enhanced thermal stability, physico-chemical characteristics, and porosity, with renewability and biodegradability features. Natural fibers can be easily modified to suit the functional requirements for biosorption by ion exchange. Graft polymerization, irradiation, plasma treatment, ultrasonic treatment, and gas-phase oxidation are some polymer modification methods. The process increases the active sites on the fiber polymer to adsorb the target metal ion, chemical, or color. Cationic dye containing positive sites requires negatively charged ions on biopolymers for their removal from wastewater [78]. Agro-waste such as banana pseudostems, kapok, milkweed, tellicherry bark, and typha fibers are reported to adsorb heavy metals such as lead, cadmium, and mercury and color from contaminated effluents and polluted soil [78–82]. Plant biomaterials assisting in heavy metal and color removal are presented in Chapter 5.

4.3.4 Nano and Microfibril Cellulose

Renewability, availability, and biodegradability are some of the most common advantages of cellulose. Nanocellulose biomaterial is nontoxic and has biomedical and food packaging applications [83]. Plant cellulose is an important alternative material for bone tissues as it biologically mimics the bone extracellular matrix, integrates with bones, and supports their growth and the growth of adjacent bones. Plant celluloses are biocompatible with good tensile strength and suppress bacterial growth. Hydroxyapatite-coated kenaf is considered a potential biomaterial for bone growth [84]. Plant cellulose nanofiber aerogels are used in wound dressing, medical, and hygiene applications [85]. Three types of nano-cellulosic structure are (1) cellulose nanocrystals (CNCs), (2) cellulose nanofibrils (CNFs), and (3) bacterial cellulose (BC). Three methods of cellulose nanofibril preparation are mechanical (homogenization, ball milling, grinding), enzymatic and chemical (hydrolysis, quaternization, TEMPO oxidation), and combinations of mechanical and chemical treatments. Steam explosion under pressure is a promising pretreatment to fractionalize lignocellulosic biomass [83, 86]. The most common plants used to make nanocellulose are rice straw, wheat straw, husks, bagasse, hemp banana leaf, soybean hulls, palm oil residue, pineapples, and flax straw. The process of nano- and microfibril extraction from biofibers is done by removing hemicellulose, lignin, and pectin from cellulosic biomass. TEMPO (2,2,6,6-tetramethylpiperidine 1-oxyl) oxidation is an energy-efficient mild mechanical method to form nanocellulose. The same method yields nanocellulose of various sizes based on source of the cellulose. Nanocellulose from cotton and wood will be in the range of 3–5 nm, whereas tunicin and bacterial sources yield 3–20 and 3–100 nm sizes, respectively [87]. Cellulose nanocrystal are crystalline with length 100–600 nm, whereas cellulose nanofibrils consist of amorphous and crystalline both regions with length up to 1000 nm. CNC and CNF have antimicrobial applications and add strength and thermal stability to reinforce the polymer matrix in heavy metal absorption [88, 89]. Composites made of nanocrystalline cellulose exhibit excellent thermal and mechanical properties along with reduction in water vapor permeability [83]. Microcrystalline cellulose is used in gels, foams, drugs, food, composites, cosmetics, paper, and various biomedical applications (Figure 4.1) [90].

4.4 AGRO-RESIDUE FIBERS

Agro-residues are the discarded parts of harvested crops and by-products of agricultural industrial processing. Agricultural residue may be waste for food grain producers and processors and discarded either in the fields or dumped from processing units, but it has immense potential in the field of technical textiles. The agro-residue fibers presented are banana, coir, corn, cotton waste [91], oil flax [92–94], kapok [95], tires [96], lotus [97, 98], milkweed [95, 99], pineapple [24], rice straw [23, 100], sugar palm [101], tellicherry bark [102], and typha [77]. The fibers and their respective applications and crop growing statistics are presented in Tables 4.8 and 4.9, respectively. The plants/trees and their respective fibers are

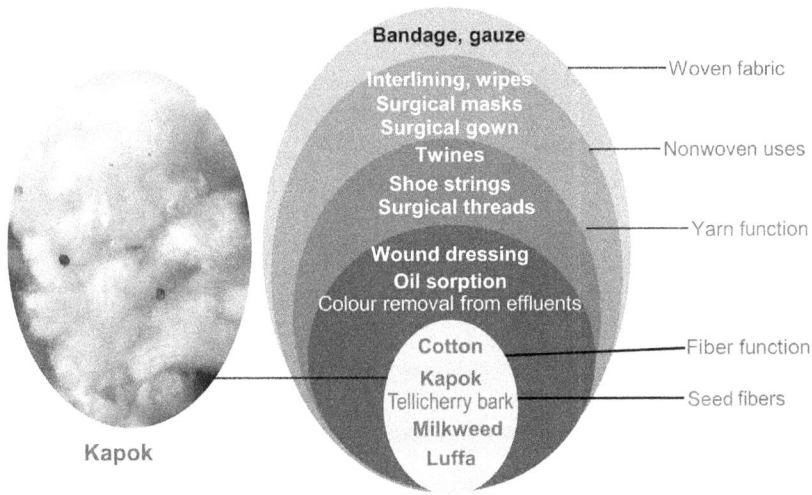

FIGURE 4.1 Biomedical applications of agro-seed fibers.

TABLE 4.8
Technical Textile Applications of Natural Fibers

Natural Fibers (Obtained from)	Botanical Name	Technical Applications	Reference
Coir	*Cocos nucifera*	Composite, coir peat, doormats, rugs	[7, 26]
Cotton (seed, stem)	*Gossypium hirsutum*	Oil spill cleanup, thermal and sound insulation	[80, 91]
Flax/linseed (stem)	*Linum usitatissimum*	Biocomposites, bags, paper, ropes, carpet, bioenergy	[92, 93, 94]
Jute (stem)	*Corchorus olitorius, Corchorus capsularis*	Composites, geotextiles, handicrafts, bioenergy	[4]
Kapok (seed)	*Ceiba pentandra*	Oil spill cleanup, thermal and sound insulation	[91, 95]
Kenaf (stem)	*Hibiscus cannabinus,*	Reinforced composites	[96]
Lotus (peduncle)	*Nelumbo nucifera*	Artificial muscles, roti wrappings, food packaging	[7, 97, 98]
Milkweed, vegan silk, vegan wool (seed, stem)	*Calotropis gigantea, Calotropis procera*	Oil spill cleanup, thermal and sound insulation	[95, 99]
Pineapple/ananas (leaf)	*Ananas comosus*	Composites, thermal insulation	[25, 52]
Rice (stem)	*Oryza sativa, Oryza glaberrima*	Brick mixing, animal bedding, composites, cement setting	[23, 100]
Roselle (stem)	*Hibiscus sabdariffa*	Biocomposites	[18]
Sisal (leaf)	*Agave sisalana*	Ropes, twines	[4, 58]

(Continued)

TABLE 4.8 (Continued)

Natural Fibers (Obtained from)	Botanical Name	Technical Applications	Reference
Sugar palm	*Arenga pinnata*	Geotextiles	[101]
Tellicherry bark (seed)	*Holarrhena pubescens*	Medical segment, face masks	[102]
Typha, cattail, bulrush (leaf, seed)	*Typha domingensis, Typha angustifolia*	Composites, reinforced particle board, paper	[77]

TABLE 4.9
Global production of Fiber Crops [FAOSTAT 2021]

Crop	Production (Million Metric Tons)	Top Producer Country	Production (Tons)
Banana	124	India	29124000.00
Coir	63.7	Indonesia	591440.97
Corn	1210	USA	348.75
Linseed/oil flax	3.34	Russia	1.3
Hemp tow waste	287318.37	France	0.0001
Jute	3457634.23	India	1.72
Kapok	78674	Indonesia	0.19
Manila fiber	107309.23	Philippines	0.07
Pineapple	28650000	Costa Rica	2.94

presented in Figures 4.2 and 4.3. Fibers are explained with respect to their potential in the section on technical textiles.

4.4.1 BANANA

Banana, valued for its fruit, is the seventh-ranked crop in the world. Fruit harvesting makes the succulent pseudostems of the plant go to waste. A 1-ha banana farm generates 220 tons of agro-waste containing pseudostems, leaves, sheaths, rachis, and bulbs. A banana fiber decorticator separates the fibers from the fleshy parts of the pseudostem. Decorticated fiber is water retted for prepreg removal. The physical characteristics of banana fiber are similar to those of roselle fiber. The diameter ranges between 3 and 7 tex. The low density (1.35 g cm^{-3}) and multiluminar structure of the fiber make it an alternative material for thermal insulation applications [103–105]. Ultraviolet protection, moisture absorption, and antioxidant properties make it an ideal fiber for paper and packaging material. Banana fabric reinforcement in composites using the hand layup method exhibited superior flexural and tensile strength [106]. Developed composites have found application in

FIGURE 4.2 Fiber-yielding plants: (a) wild flax, (b) kapok, (c) corn, (d) milkweed, (e) tellicherry bark, (f) typha, (g) coconut.

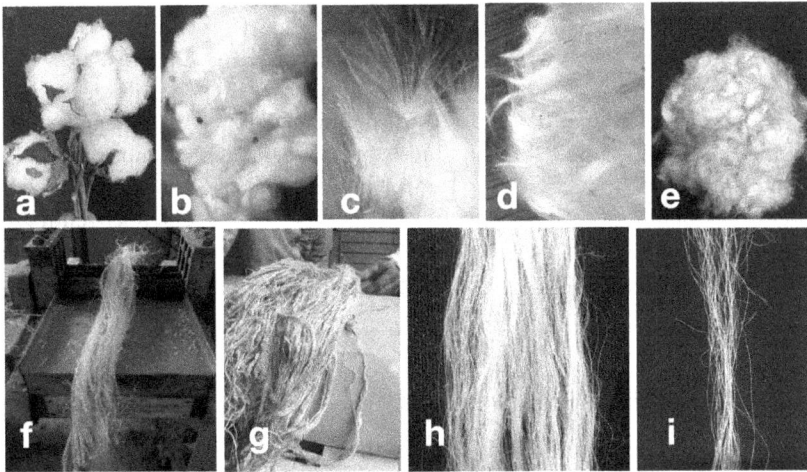

FIGURE 4.3 Plant fibers: (a) cotton, (b) kapok, (c) milkweed, (d) tellicherry bark, (e) typha seed, (f) flax, (g) banana, (h) lotus, (i) typha leaf.

acoustics and building components. Fiber compatibility with the polymer matrix is improved by chemical and enzymatic treatments. Alkali, acid, enzyme, and TEMPO-mediated treatment improves the fiber appearance by removing impurities and also makes the fiber thermally stable [107–109]. A silane and acetylation reaction imparts hydrophobicity in the fiber by replacing the -OH group of the fiber. The resultant fiber is hydrophobic, exhibiting better adhesion with resin. Solvent-free acetylation using acetic anhydride and N-bromosuccinimide improves the oil absorbance of the fiber and placeable for controlling oil spills in marine ecosystems [110]. Banana fiber is used in furnishing material, shoes, rugs, ropes [111, 112], soil enhancers [113], supercapacitors [114], and utility handicrafts. Banana fiber needle-punched nonwoven fabrics have good thermal and sound insulation properties [103, 104]. A green composite was developed with banana fiber and tamarind seed gum as a matrix with high tensile strength and flame retardancy [115]. Banana fiber papers possess high tensile, bursting, and tearing strength. The brightness of the paper increases with bio-bleaching of the pulp using xylanase enzyme [116]. The process of cellulose microfibril extraction and resultant microfiber size is presented in Table 4.10. Banana fiber ground to fine fiber assists in removing 70% of lead, cadmium ions, and mercury [79, 117]. The use of banana pseudostems and allied plant biomass in the handmade paper industry at the local level could mitigate pollution problems caused by the conventional paper industry. Large amounts of discarded banana pseudostems are a potential material for making nanocellulose for use as a polymer matrix and value-added green products [88].

4.4.2 COIR

Coir, waste from the coconut industry, is an abundantly available agro-fiber obtained from the mesocarp of the coconut palm drupe. Coir fiber consists of 200 to 300 elementary fibers, lumen, and a central lacuna. The fiber is hollow with 20 to 30% porosity, a circular cross-section, and a high microfibrillar angle of 45° [118]. The individual fiber has 10–20-μm diameter and 1-mm length [119]. The fiber is strong, coarse, thick, and resistant to degradation compared to other commercial fibers [26, 118]. Coir has the highest toughness tension compared to other natural fibers and therefore is ideal for monofilament fiber-reinforced cementitious composites. Low carbon footprint, high elongation, and excellent strength and toughness make coir a promising fiber to replace, glass, steel, and

TABLE 4.10
Nanocellulose Obtained from Banana Agro-Waste and Its Characteristics

Banana Parts	Microfibril Extraction Process	Fiber Size	Fiber Length	Applications	Reference
Pseudostem	Acid-catalyzed hydrolysis	500 μm	> 500 μm	Fiber-reinforced polymer	[89, 108]
Rachis	Peroxide homogenization under acid medium	5 nm	500 nm–1 μm	Reinforcing filler in nanocomposite fabrication	[90]
Pseudostem	TEMPO-mediated oxidation + high-intensity ultrasound treatment	3–10 nm	140–690 nm	Polymer matrix	[88]
Pseudostem	TEMPO-mediated oxidation	7–35 nm	> 500 nm	Packaging material	[109]

carbon fibers in reinforcement of cement-based composites. Coir fiber-reinforced particle boards require less energy than humanmade fiber-reinforced material [59]. A higher strain to failure increases fiber toughness in reinforced composites. The low density of coir improves the compressive, impact, and tensile strength, along with thermal and sound insulation of the cementitious matrix. Coir with a polyester matrix demonstrates excellent (100%) sound absorption at 1700 Hz. Fibrous coir of 20 mm thickness with 0.1978 g cm^{-3} density absorbs 90% of sound at a 3000-Hz frequency. More fibers and higher density and thickness increase the absorption coefficient at lower frequencies [120]. Coir has been judged the most suitable fiber for sound insulation properties compared to cotton and sugarcane fibers [57]. The fiber has diverse applications, including biocomposites, ropes, rugs, coir peat, and bio-pads (Figure 4.4) for plants [7, 26, 118].

4.4.3 Corn

Maize is a major crop grown the world over for its edible golden grains. It is an all-season crop, sown thrice a year depending upon the variety. Crop harvesting is done to collect the corncobs containing seed kernels [121, 122]. The rest of the plant parts, such as stalks, leaves, and husks, are discarded in the field itself. Common practice is to burn the leftover residue in the fields. Stalk, leaf, and husk contain fibers of various grades. Leaves and stalks are water retted to obtain fibers [123]. Corn-growing statistics (Table 4.9) show large-scale availability of corn stover. One ha of corn field generates 0.12 tons of leaf fibers. Corn leaf fibers are a potential source of paintbrushes, doormats, packaging, baskets (Figure 4.5), and sound insulation material. Corn husk ash can be used as a replacement for cement [59].

FIGURE 4.4 (a) Ornamental vines thriving on coir pads, (b) peat (c) scanning electron microscope image of coir extracted from coconut trunk.

FIGURE 4.5   Handicraft utility items made of corn fiber.

4.4.4 Cotton: Pre- and Post-Consumer Waste

The ingenious morphological structure, moisture absorbance, moisture management, and dyeability of cotton fiber are unmatched and yet to be mimicked in
anthropogenic fibers [1]. Seventy-five percent of pre-consumer waste is recycled
by the industry itself, mainly for technical textile segments [6]. Pre-consumer
waste is utilized in nonwovens and spinning coarse thread for mattresses, automobiles, paper, furnishings, and composites. Cellulose and protein fiber lint mixed
with farm waste display water-holding capacity, requiring less irrigation [124].
The incorporation of bio-fibers in concrete will provide sustainable housing solutions. In this regard, cotton lint, industrial waste, and recycled cotton may contribute to the construction industry in big way, supporting the principles of circular
economy [6, 125]. Fibers extracted from pre-consumer waste and waste cotton
fabric were reinforced for composite fabrication suitable for dashboards and furniture [37]. A rag-tearing machine was used to open the fibers from the fabric.
The fibers considered shoddy were carded prior to epoxy resin application. Epoxy
resin and curing agent composition was in the ratio of 100:20. Compression molding (Table 4.7) was used to fabricate a shoddy-epoxy composite. Composites were
thermally stable and had a lower water diffusion coefficient than oak, pine, and lin-
den wood. Mechanical properties (49.47–83.75 MPa), elongation (5.25–6.96%),
and impact strength (2.13–7.62 kj m^{-2}) increased with increasing amounts of fiber
fraction in the composite. Developed composites, comparable to commercial
wood in mechanical properties, can be a substitute for timber [37]. Cotton stalk,
which is considered waste and discarded, is a source of fibers rich in cellulose
(79%), with a modulus of 144 g/d. The fibers are a potential source of reinforced
composites and biofuel. Cotton stalk fiber-polypropylene composites have tensile,
flexural, and impact resistance similar to jute-polypropylene composites [126].

Shredded textile waste mixed with cement and water produces a composite with wood characteristics that can be nailed and cut into shapes. The developed composite was lightweight, fire resistant, and a low-cost alternative to concrete blocks. The sturdy material has good potential in walls, ceilings, and wooden boards as a substitute for concrete [127]. Waste cotton clothing and denim were used to procure ethanol via an enzymatic hydrolysis and fermentation reaction. The same process can be utilized to obtain ethanol from cotton litter [128].

4.4.5 KAPOK FIBER

Light and fluffy kapok fiber is highly hydrophobic. Kapok fiber has a high wax content, and 77% of its total fiber mass is lumen. The waxy surface of kapok is water repellent (contact angle > 90). Natural fibers such as cotton, kapok, and milkweed are termed conventional fibers in the field of technical textiles. Kapok and milkweed fibers are hollow, which makes them suitable for oil sorption. The presence of lumen in cotton, kapok, flax, and milkweed accelerates the capillary action during oil absorption. Sorbents recover spilled oil by absorbing (soaking) and adsorbing (storing). Adsorption of oil using fiber attracts and stores oil layer by layer by binding physically (Van der Waals forces) and chemically (electron exchange). Absorption and adsorption together hold the absorbed liquid in surface pores. Sorbents made of fibers help to retain the spilled oil and subsequently in its recovery, recycling, or disposal mechanism. Oil sorption by cotton and kapok is 1.5–3.0 times greater than that by polypropylene fibers [91]. Nonwovens made of milkweed, cotton, and polypropylene blends are able to completely remove oil from water [80]. The higher wax content and hollow structure of natural fibers make them efficient for oil sorption; 20–30% of oil sorption by cotton fibers is due to the wax content of the cotton. Natural fibers are therefore the best materials to tackle the problem of oil spillage. The oil sorption capacity of 1 g textile fibers is presented in Table 4.11 [91]. Cellulosic fibers from waste textiles have a high rate of heavy metal absorption [129]. End-of-life management of natural fibers does not pose any environmental threat, as they easily biodegrade in water as well as landfill and do not generate toxic fumes while burning in biopower plants. Pyrolysis of waste generates steam, which is safe to utilize in sterilizing reusable boxes used to transport medical waste. The oil sorption potential of natural fibers presented in Table 4.11 is in sync with their wax content. The higher value of polypropylene is due to the porosity of the fine fibers, leading to higher capillary bridges between fibers [91]. Kapok is composed of a comparatively high amount of lignin and wax. Higher lignin and wax content together with the hollow structure of the fiber contribute to metal ion and oil sorption [91, 95]. The higher lignin content (20.73%) of kapok plays a key role in repelling water as well as a higher glass transition value (142 °C). Kapok blended with cotton at a ratio of 2:3 can be spun. Traditional technical applications of kapok are in thermal insulation; achieving desired soundproof standards; and filling mattresses, upholstery, pillows, and stuffed toys [130, 131].

TABLE 4.11

Oil Absorption by Textile Fibers [91, 99, 131]

Fiber	Diameter (μm)	Density (g/cm³)	Wax (%)	Oil Sorption Capacity (g/g)
Cotton	21	1.54	2.9	50
Kapok	23	0.29	5.31	60
Milkweed	29	0.97	2	40
Polypropylene	19	0.905	-	80
Blend of previous four fibers	-	-	-	40.16
Flax (untreated)	35–85	1.5	1.4	13.75
Flax (acylated)	60–72			24.57
Flax (microwaved)	60–72			17.42
Waste tire rubber aerogel	-	0.020–0.090	-	25

4.4.6 LOTUS FIBER

One of the most ancient fibers, a symbol of courage and the highest survival instinct, lotus fiber is being recognized in the current scenario of sustainable sourcing [6, 97]. The finest microfiber is obtained from plant peduncles. Lotus peduncles are considered waste in large parts of the world. The plant peduncles biodegrade in water bodies post-flower blooming or lie outside as a drive to clean water bodies. The fiber has a unique shape and elliptical to oval cross-section without lumen. It has a cellulose I crystalline structure and a promising future in the industrial segment [132]. The fine fiber has a naturally coiled structure that results in unique properties like natural wrinkle resistance in garments [98]. Technical uses of lotus fiber are in biogenics as artificial muscles. Lotus fiber muscles in human body are capable of bending, twisting, and rotating due to the natural twisted configuration of the fiber. Lotus fiber muscle has 38% tensile strength and 450 j/kg work potential during contraction. The contraction is 56 times more than the natural skeletal muscles and higher than any other natural fiber muscles. The rotation speed reaches up to 200 rpm, with a torsional stroke of 200°/mm. Diverse applications of lotus fiber artificial muscles are in weight lifting by artificial limbs and smart textiles [133]. Composite film fabrication using bismuth and lotus fiber is placeable in photocatalytic activity under light irradiation. The film is a potential sewage disposal and phytoremediation tool able to degrade methylene blue up to 94% in 270 min [134]. Lotus fiber processing is done by hand without chemical intervention from the fiber to fabric stage. Therefore, lotus fiber fabric's potential in the food packaging industry, such as roti wraps, bread, and tortilla packing, is promising [98].

4.4.7 MILKWEED FIBER

Fine, soft, and lustrous milkweed (*Calotropis gigantea* and *Calotropis procera*) is a single-cell fiber obtained from the Apocynaceae family. Seed fiber measuring

30–40 mm and 20 μm in length and diameter, respectively with 56% crystallinity and significant mechanical strength [130]. The high absorbency and moisture transmission of milkweed fiber are due to the presence of hollow channels along its length. The porous structure of the fiber makes it suitable for sound absorbency and thermal insulation. The noise absorption coefficient (NAC) of 100% milkweed fiber needle-punched nonwoven web is in the range of 0.582–0.989. The NAC value of milkweed fiber is more than those of hollow polyester (335–383) and blends of milkweed and polyester. Soft and flexible nonwovens made of milkweed fiber used in building and automobile construction will help in countering noise pollution [7, 99]. Heavy metal–contaminated effluent remediation using lignocellulosic fiber filters is gaining popularity, as they are abundantly available, low cost, and environmentally sustainable. Oil sorbents made of natural fibers have an edge over synthetic fibers due to better cleanup and end-of-life management potential [91]. Milkweed nonwoven fabric (0.1 g) is useful in removing heavy metal ion concentrations (20 mg/l) of Pb^{2+} and Ni^{2+} from an aqueous solution of 50 ml/l. The tannin and lignins present in cellulosic fibers are active sites for absorbing heavy metal ions. The ion exchange capacity of plant and animal fibers is due to the presence of cellulose, lignin, and protein. The slightly acidic pH of 6 is optimal for metal ion adsorption. At this pH, Pb^{2+} and Ni^{2+} ion removal at 26 °C was 96.16% in 90 min and 65.58% in 60 min, respectively. pH values below this and beyond hamper metal ion adsorption capacity by limiting the active H^+ ion concentration. An increase in adsorption time beyond the optimum (60–90 min) caused noticeable desorption of metal ions. Active sites on the surface of milkweed also decrease with a higher initial concentration of metal ions, limiting metal ion adsorption [95]. Alternative industrial applications of milkweed fibers are in geotextiles, filtration, and medical segments [99]. Milkweed stem fibers have found applications in biocomposite materials with high physical performance [25].

4.4.8   OIL FLAX AND ALLIED OLEAGINOUS FIBERS

Global production of oil flax was 3,339,000 tons in an area of 4142.45 h in fiscal year 2021. One ha of flax cultivation yields 2500 m² technical textiles in a functional unit per year [135]. The production data and cultivation area may provide thousands of clusters in the world where oil flax fiber production units can be operational, benefitting the farmers and sustaining the crop. Besides fiber, shives or boon, which is 75% of the stem, is also a potential source of composite fabrication [94, 136, 137].

Oil flax, fiber flax, and dual-purpose material possess comparable physico-thermo-chemical properties and are suitable reinforcement materials for polymers [94, 137, 138]. Oil, oleaginous, or linseed flax is grown for oil purposes. The plant height is lower than the other two types, but it contains fibers suitable for technical applications. Flax fiber has three elementary layers consisting of different cellulose microfibrils and a hollow lumen. Plant phenotypes, fiber retting, and physical characteristics have been the subject of intensive investigation in earlier studies

[94, 139–142]. Table 4.12 presents the physico-chemical characteristics and production potential of flax fiber. The primary cell wall consists of amorphous pectin, lignin, and hemicellulose, with an microfibrillar angle of 5.8 to 7.3°. The secondary cell wall consists of cellulose (78%) and is further divided into three sub-layers: S1, S2, and the S3 or G layer (Figure 4.6). S1 microfibrils have a Z-twist orientation with an uncertain angle, whereas S2 has a microfibrillar angle of 6 to 11°. The S3 or G layer, placed next to the central lumen, is considered the gelatinous muscles of the fiber, which contract and change orientation in response to mechanical stress, supporting the plant. The secondary cell wall and microfibrillar angle are important factors responsible for the mechanical properties of the fibers

TABLE 4.12
Fiber Flax Production Potential

Properties	India*	USA**	European***
Plant height (mm)	< 500	1530–1760	1295–1466
Seed production (t/h)	1.38	0.4–1.4	1.99–4.4
Fiber production (t/h)	0.4	-	1.22–2.67
Straw yield (t/h)	2.0		2.83–3.97
Fineness (tex)	2.7	1.5	2.65
Tenacity (cN/tex)	23.45–29.9 g/tex	27.0–33.0 g/tex	38.0 g/tex (69.5)
Elongation (%)	1.98	1.8–2.3	2.46–5.53 (5.2)

Notes: * *Shekhar* (oil flax) phenotypes, *Tiara* properties [93, 94, 147]
**Ariane* [94, 140]
***European line flax phenotypes, *Viola, Modran*—physical properties [94, 141, 147]

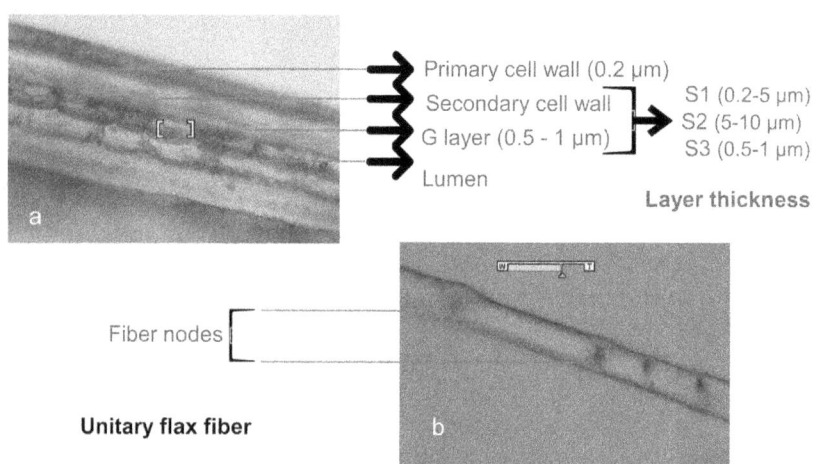

FIGURE 4.6 Flax fiber microscope image exhibiting fiber nodes, central lumen, primary and secondary cell wall.

and consequently the biocomposites made thereof [130, 143, 144]. The average value of the flax fiber microfibril angle is 5° in the case of the *Bolchoi* variety (diameter 2 μm) cultivated in Normandy. The microfibrillar angle of flax fiber increases under wet conditions [144]. A lower angle of microfibrils is linked with a higher Young's modulus, yielding higher mechanical strength. In this regard, the microfibrillar angle of flax (5°) is lower than those of hemp (6.2°), pineapple (6–12°), jute (8.1°), banana (11–12°), sisal (10–22°), cotton (20–30°), and coir (39–49°). This explains why flax, hemp, pineapple, and sisal are preferred fibers for engineering composites, whereas the excellent properties of cotton continue to be exploited for fine and comfortable apparel [130]. The presence of nodes (Figure 4.6), which are circumferential dislocations in fiber, display a higher microfibrillar angle than the entire cell wall width [143]. A high Young's modulus, strength, low cost, low density, and lower environmental impact make flax a promising fiber to replace glass fiber in composite manufacturing in aerospace, sports, and the design industry [144, 145]. Flax-based technical textiles include greener automobiles and everyday products. Fiber-reinforced composites, anti-sore bedding for patients, bicycles, shock-absorbent tennis rackets, paper, bags, wipes, wicks, sanitary napkins, briquettes, ropes, shipping cords, mats, fillers for saddle pads, car components, handicrafts, and oil sorption (Table 4.11) are some of the reported technical textile products made by incorporating flax fibers [54, 92, 94, 135, 146–148]. *Naturalis historia*, penned in 77 CE by Pliny, describes the use of fiber tow as bags, ropes, wipes, and wicks [149]. The historical, cultural, and nutritional significance of flax has convinced farmers to continue with its production through generations. Over the years, several minor oil beans like Mahua (*Mahua longifolia*), conophore (*Plukenetia conophora*), honey locust (*Gleditsia triacanthos*), niger (*Guizotia abyssinica*; Hindi: *ramtil*), shea nut, teff seeds, amaranth seeds (*Amaranthus caudatus*; Hindi: *ramdana*), and poppy seeds have become extinct in larger parts of India, but flax survived as a source of raw material for the food, fiber, and furnishing industries. Perhaps the ancient world was aware of the numerous advantages of flax oil and fibers such that flax cultivation continued throughout [92, 140, 150–152]. Castor oil plant stems, mustard straw, canola straw, palm, lemongrass, and datura are allied oleaginous crops useful in fiber production with good thermal resistance and stability for door panels, paper, packaging, insulators, nonwovens, and electric goods [153–155]. The advantages of using oleaginous crops for technical textiles are their hollow structure, low density, and excellent mechanical properties. Canola (*Brassica napus* L.) fiber density is lower than that of cotton, and it requires less irrigation compared to cotton crops. Plant biomass is suitable for nonwovens, eco-composites, and technical textiles [153]. Oil flax fiber is useful in fabricating bedding with anti-bedsore properties [136, 149, 151].

4.4.9 PINEAPPLE FIBER

Pineapple leaf fiber has a high initial modulus (570–700 cN/tex), which makes it suitable for use in industrial textiles such as conveyor belt cords and lightweight duck cloth. Fiber composites are used in automobiles and railway coaches. Pineapple leaf fiber production potential in the three countries with the highest cultivation area is approximately 240,000 (Nigeria), 133,000 (India), and 103,000 tons (Thailand). The fibers have great potential in technical and non-technical applications [24]. Common commercial thermal insulators used in buildings are glass fiber, metal wool, rock wool, and polystyrene foams. Synthetic fibrous metal mats on walls and ceiling open spaces are popular but have ill effects on the human body and environment [52]. Particle boards made with pineapple leaf fiber mixed with a natural rubber latex binder serve as a good alternative to synthetic fibrous thermal insulators (Table 4.8). Fiber treatment with 18% sodium hydroxide enhances its strength and elongation and imparts crispiness. Various uses of pineapple fiber fabrics are in ropes, bags, table linens, interior design products, and mats that are lightweight and stiff [24].

4.4.10 RICE STRAW

Rice straw contains strong and spinnable cellulosic fibers. Rice grains are a staple food the world over. Rice straw and rice husk are agro-residue that remains in the paddy fields post-procurement of grains for human consumption. The sourcing of rice straw, among the most widely available residues, as bales is one of the cheapest options at 100 USD/tons of cellulose [3]. Traditional uses of rice straw are in cattle feed and to provide cushions for animal bedding. Rice husk is utilized as a filler in cow dung cakes, bricks, and utensil cleansers. Rice straw used in building construction provides comfort and thermal stability, especially in tropical regions where solar radiation exists for a longer duration with high intensity throughout the year. Rice straw is mixed with concrete, which prevents the building from cracking due to high thermal fluctuation [23]. Composites made of rice straw fibers have suitable sound insulation material in wooden constructions. The bending strength of rice straw particle board is comparable to that of wooden particle board [156]. Heat transfer of rice husk ash (burned at 600 °C)-sand-cement brick reduces by 46 W compared to traditional clay brick. Thus, rice straw and ash mixed in concrete roof tiles improve insulation properties and level of comfort besides being a low-cost housing solution [23, 100]. A study on agro residue–reinforced cement composites utilizing coir, corn cobs, bagasse, durian peels, and palm leaves reveals the lowest thermal conductivity for rice husk insulation boards, followed by bagasse, coir, durian peels, corn cobs, and palm leaves. The lowest density was observed in bagasse, followed by rice husk, corn cob, coir, durian peel, and palm leaf insulation boards. Rice husk ash and a mineral admixture are a good replacement for cement as they improve strength and decrease water absorption by the concrete [100].

4.4.11 SUGAR PALM FIBER

Sugar palm (*Arenga pinnata*), belonging to the Palmae family, is a source of naturally available woven fabric. Woven fiber mats consist of lignocellulosic fiber (*ijuk* fiber) useful in nano-cellulose fiber (NCF) formation. Common vernacular names of sugar palm are Arenga palm, black fiber palm, *taad* (Hindi), *gomuti, kaong, aren, enau*, and *irok*. Palms grow to a height of 20–30 m. The tall straight trunk is used to make wooden spatula and furniture. Sugar palm fiber collection requires removing the fiber mats wrapped around the palm stalk. The separated fibers are washed with water to remove dirt and other impurities and subsequently dried for two weeks [16, 101, 157]. Traditional uses of sugar palm fiber are in carpets, fillers, brooms, ship cordage, doormats, water filters, paintbrushes, fish breeding nets, and thatch roofs [158]. The average fiber diameter and cellulose content of sugar palm fiber are 212 µm and 43.88%, respectively. The fiber diameter is reduced by acidic (94.49 µm) and alkali (11.87 µm) treatment. Mechanical homogenization further refines the sugar palm fiber (3.9 µm) and improves the cellulose content from 43.88 to 88.79%. The process noticeably increases holocellulose from 51.12 to 94.64% and reduces lignin from 33.24 to 0.04% [159]. The cellulose, lignin, and ash content of the fiber varies on the basis of palm age and parts along its length. Variation in the chemical constituents of the fiber is depicted in Table 4.13. The chemical content of the fiber is unchanged above the 15 m height. Higher ash content at lower parts of the palm is due to impurities embedded in the fiber, mainly silica [101]. Chemical and mechanical treatment of sugar palm fibers removes lignin, inter-fibril hydrogen bonds, and the hemicellulose content of the fibers. Efficient refining and delignification of sugar palm fibers results in an improved crystallinity index (81.2%) and lower degree of polymerization (289) [159]. Naturally woven fiber mats are highly durable and resistant to seawater. Fibers collected from different plant heights (1 to 15 m) were studied in detail. The tensile modulus was found to be highest (3.37 GPa) in fibers collected from 13 m plant height. Fiber elongation and toughness was maximum at 7 (28.32%) and 11 m (52.46 MJ/m^3) height, respectively. Sugar palm fiber uses in the area of technical textiles include geotextiles for road construction, prevention of soil erosion, composites, biopolymers, and underwater and ground cables. The higher silica content of the fibers makes them suitable for use in composites as a thermal insulator. Fiber treatment with seawater fibrillates the fibers and improves interfacial bonding with the composite matrix, resulting in improvement of the mechanical properties compared to the untreated fiber composite. The low density of sugar palm (1.22–1.26 kg/m^3) makes the composite product lightweight in comparison to the traditional glass fiber (2.55 kg/m^3)–filled composite. A comparison of reinforced composites made of an inner layer using sugar palm woven fabric, long fibers, and short fibers showed the tensile and flexural

TABLE 4.13

Sugar Palm Fiber Chemical Composition at Different Plant Heights [101]

Fiber	Fiber Length	Hemicellulose (%)	Lignin (%)	Ash (%)	Moisture (%)
Sugar palm	37.3	4.71	17.93	30.92	5.36
3 m	49.36	6.11	18.9	14.04	8.64
5–15 m	53.41–56.8	7.36–7.93	20.45–24.92	2.06–5.84	7.72–8.7

performance was highest in the composites containing sugar palm woven fabric, followed by long and short fibers, respectively. Fibers from mature live palms from 1 m height are suitable for reinforcing in a composite, as they are superior in strength, elongation, toughness, and thermal stability. The four stages of thermal degradation, in order of moisture, are (45–125) >> hemicellulose (210–300) >> cellulose >> lignin >> ash. Crystalline cellulose degradation starts at 300 °C, and it completely decomposes at 340 °C. Lignin degradation, which starts at 160 °C, continues to 900 °C. Inorganic component ash remains longer and decomposes at a high temperature of 1723 °C due to presence of silicon dioxide [101].

4.4.12 TELLICHERRY BARK FIBER

Tellicherry bark fiber is obtained from the tree seedpods (Figures 4.2 and 4.3) [102]. The fibers are useful for heavy metal and color removal from dye effluents. Defatted and delignified tellicherry bark fibers were copolymerized with acrylonitrile prior to color removal treatment. Modified tellicherry bark fibers were capable of removing 99% of malachite green from dye effluents within a time frame of 165 min. Adsorbent dose was 500 mg/50 ml of dye concentration. The adsorption process by the treated fibers followed both macro- and micropore diffusion mechanisms [78]. Tellicherry bark–cotton blended yarns were knitted into plain fabric. The comfort properties of the blended knits were found to be better than pure cotton knitted fabric [102]. The blended knits were fabricated to make face masks and medicated linings of plaster bandages (Figure 4.7).

4.4.13 TYPHA FIBER

Tall typha leaves (1.4–3.0 m) are strong, spongy, and smooth, with a composite structure to withstand harsh environmental stress. The most common names of

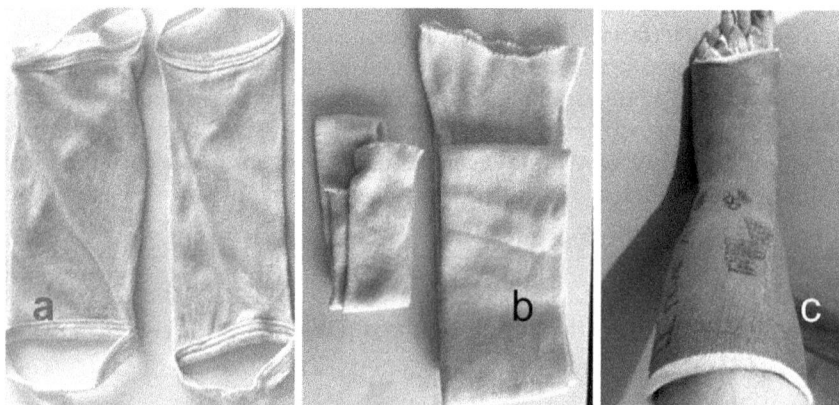

FIGURE 4.7 Tellicherry and cotton yarn knits: (a) face masks, (b) orthopedic stockings, (c) stocking linings on orthopedic bandages.

typha are cattail, bulrush, reedmace, and elephant grass. Plant leaves are pounded prior to water retting, followed by washing in running water, untying, and extraction of loose fibers by hand [77]. Fiber exhibits high cellulose (68.13%), tenacity (38.58 g/tex), and low density (1.34) at three months of plant age. The advantages of typha fiber panel boards are high load bearing capacity and superior sound insulation. Plant leaves are utilized in particle boards for building construction, roof thatching ropes, and baskets. High aspect ratio, thermal stability, and strength make typha fiber a potential source of reinforcement in polymer matrices for the automotive and aviation industries. Higher-density particle board exhibited high strength, modulus of rupture, modulus of elasticity, moisture content, and thermal conductivity. Thermal insulation was found to be greater in typha fiber boards, with lower-density boards compared to high-density boards with similar thickness. Boards with lower density have voids that make them lighter in density and lower in thermal conductivity compared to high-density boards with identical thickness [56, 77]. Wetland and macrophyte plants such as typha and water hyacinth absorb heavy metals, salts, and chemicals from wastewater, making the soil and water toxin free. Plant-generated biofuels fertilize the soil with nitrogen and enrich the surroundings [7, 77].

4.5 PHYSICAL CHARACTERISTICS OF AGRO-RESIDUE FIBERS

The physical properties of agro-residue and various other commercial plant fibers are presented in Table 4.14. The quality of agro-residue bast fibers is influenced by environmental factors, post-harvest handling, and retting efficiency [92, 94, 160–163]. Fiber diameter, density, and lignin increase with plant age,

TABLE 4.14
Comparative Chart of Physical Attributes of Agro-Residue Fibers

Fiber	Fineness (d)	Length (mm)	Tenacity (g/d))	Elongation (%)	Reference
Banana	21	600–660	2.4–3.7	1–3.5	[19, 160]
Coir	50–55	3–9	2.0	15–30	[19, 53, 118]
Corn leaf	1.9	100–120	1.0	4–5	[123]
Cotton	1.25	28.52	2.0–3.7	3.0–10.0 (Ngo)	[167]
Cotton stalk	51	80	2.9	3	[126]
Flax	19.8	353.0	2.65	2.78	[53, 93]
Kapok	0.4–0.7	20.0	1.4–1.74	1.4–4.23	[161]
Lotus	1.98	467	1.49	1.95	[97]
Milkweed (seed)	1.05	26.14	3.71	3.9	[99, 166]
Pineapple	31.5	1000–1200	0.7–3.8	5.6	[24, 162]
Rice straw	27	2.5–8.0	3.45	2.19	[19, 23]
Sugar palm		1.19		19.6	[157]
Tellicherry bark	1.6	29.55	1.88	2.57	[102]
Typha stem	56.7	560.0	4.28	2.9	[77]

but tensile strength, lumen size, and moisture content decrease [18, 102]. Fiber tenacity and elongation of agro-residue fibers are in the range of 1–4.8 g/d and 1–19.6%, respectively. Durable mechanical properties make them potential fibers for limited-life geotextiles and soil erosion control applications [164]. Chemical treatment further improves the thermal stability and moisture regain value of the fiber [165–167]. Fibers extracted from plant stem are coarse but high in strength, which make them suitable for carpets, biocomposites, canvas fabric, and reinforced panels.

4.6   CHEMICAL PROPERTIES

The chemical composition of agro-residue fibers is presented in Table 4.15. Agro-residue fibers are lignocellulosic fibers containing cellulose, hemicellulose, pectin, lignin, coloring matter, fat, and wax. The fiber content influences fiber color and quality. Cellulose is the principal constituent; hemicelluloses are amorphous isotropic polysaccharides; pectin is water soluble; lignin is an isotropic non-crystalline polymer; fiber cortical cells contain coloring matter; fats and waxes are present on the fiber surface and can be removed by benzene treatment. The cellulose content and moisture absorption potential of the fiber decrease with plant ageing. Plant lignin provides thermal stability to the fiber as the lignin chars above 900 °C [18, 167]. Chemical composition influences the fiber color and quality. Higher cellulose is preferred in textiles, paper, and various fibrous applications.

TABLE 4.15

Comparison of Chemical Constituents of Agro-Residue Fibers

Fiber	Cellulose (%)	Lignin (%)	Wax (%)	Ash (%)	Moisture (%)	Density (g/cm³)	Reference
Banana	31.27–64	15.07	4.46	8.65	10–11	1.35	[86, 160]
Coir	33.2	20.5		2.7–10.2	10–12	1.2	[169]
Corn stover	38–40	7–21		3.6–7.0	9.0		[19]
Cotton	95.95	-	0.5	2	8.0	1.5–1.6	[167]
Cotton stalk	79.0	13.7			8.8		[127]
Flax	79.68	8.74	1.4	2.27	7.66	1.4–1.5	[93]
Kapok	53.40	20.73	5.31	0.54	10	0.4	[161]
Milkweed	55	18	1–2	1–2	10	0.8	[99, 166]
Pineapple	70.98	4.9	0.96	0.95	7	1.5	[162]
Rice straw	64	8	3.72	5	9.8		[19, 171, 172]
Sugar palm	43.88	33.24		4.27	5.63–8.36	1.29	[16, 101, 159]
Tellicherry bark	67.6	19.1	0.48	1.77	10.43	1.05	[102]
Typha seed	73.46	9.88	0.93	5.9	9.6	1.29	[77]
Typha stem	68.35	17.6	0.37	5.71	7.8	1.33	[77]

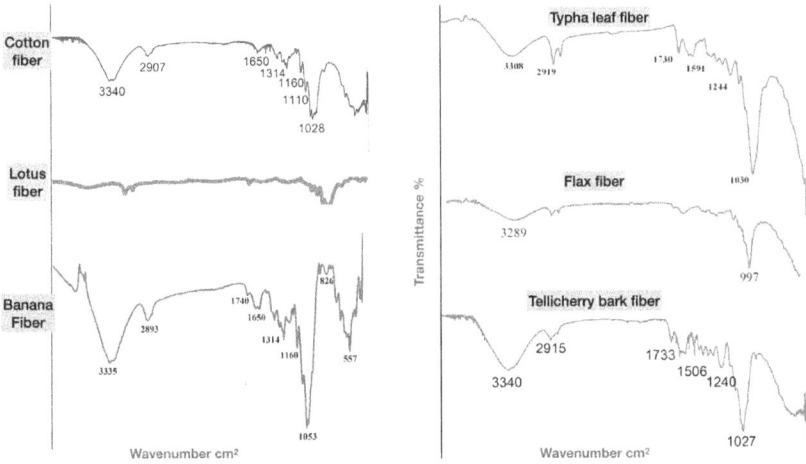

FIGURE 4.8   Fourier transform infrared spectroscopy exhibiting the presence of functional groups in agro-residue fibers.

The low density of natural fibers is useful in lightweight composite manufacturing [168–172]. Fourier transform infrared analysis (FTIR) of agro-residue fibers (Figure 4.8) shows typical bands of cellulose, hemicellulose, and lignin around 1650–1740 cm^{-1}. The range for FTIR was from 500 to 3600 cm^{-1}. Wide bands between the 3308 to 3335 cm^{-1} regions exhibit the presence of the -OH group of absorbed water. C-H stretching in alkanes is demonstrated by a small band (2893–2919 cm^{-1}). The bending vibrations of the C-H and C-O groups show the presence of aromatic rings in cellulose polysaccharides [93]. The presence of -OH and -COOH groups in the fibers' molecular chains increases the affinity for reactive dyes.

4.7   ADVANTAGES OF AGRO-RESIDUE UTILIZATION FOR FARMERS

Harvested crops and agro-industry–generated crop residue are higher in quantity than the edible grains of the respective crops. For most crops, like oil flax, the edible grains and crop residue are in the ratio of 1:3 [94]. Pineapple leaf fiber production potential in the three countries with the highest cultivation area is approximately 240,000 (Nigeria), 133,000 (India), and 103,000 tons (Thailand). The fibers have great potential in technical and non-technical applications [24, 168–177]. A 1-ha banana farm generates 220 tons of agro-waste containing pseudostems, leaves, sheaths, rachis, and bulbs [103, 106]. One ha of corn field generates 0.12 tons of leaf fibers [59]. One ha flax cultivation yields 2500 m^2 technical textiles in a functional unit per year [135]. The price of rice straw bales is 100 USD/ton of cellulose. The large humanmade cellulosic fiber-producing industry requires 1600 kilotons of dry agricultural biomass. Farmers producing staple crops only benefit from selling the edible grains. If rest of the agro-residue left on farms and agro-industries is utilized, farmers will get even greater dividends from crops. Agricultural crops are vulnerable to climate change and extreme weather affecting production yield. Soil health, degradation, plant diseases, pest attacks, and hailstorms also pose a threat to production statistics. Such natural calamities take their toll on food grain production and add to the already existing amount of crop residues. Sometimes farmers' only alternative is to plow the standing crop to prepare the land for the next crop. If agro-residue utilization is profitable for farmers, then instead of plowing the standing crop, they will have the choice to supply the harvested crop and plant biomass to the paper and anthropogenic fiber manufacturing industry. High amounts of crop residue burning are considered highly unsustainable as it causes air haze due to an increase in aerosol particles in the atmosphere. Extensive use of pesticides and groundwater extraction for improved production renders pollutes fields and drains with chemicals. Using fertilizers in crop fields contributes to the carbon footprint of the cultivated products [3]. Biocompost made from agro-residue and cow dung improves cultivation and is a safer alternative to chemical fertilizers. Toxin-free soil and saving earth and water can be achieved through agro-residue utilization in technical textiles, biocompost, and bioenergy.

4.8   CASE STUDY OF DISCARDED BIRD'S NEST FIBERS FOR AGRO-TEXTILES

A bird's nest is an ideal design using ideal material. Made of littered agro-residue, the environmentally conscious design creates no pollution or end-of-life management crisis. Weavers' birds have an unusual talent for fabricating the world's most natural home, which is an epitome of sustainability. The nests are lightweight and have an airy and open weave structure with no discernible pattern. Woven with the mother's love, affection, care, and tenacity, the nest is used for only one season to take care of the young chicks. By the next season, a newly woven house awaits the newly born chicks. The nest is woven with safety precautions for young chicks and during hatching. Parts of the nest include an entrance, which is like an open window, and a safety railing inside that protects the hatchling from falling (Figure 4.9). Abandoned nests were utilized in the field of agro-textiles as seedling sleeves (Figure 4.10) and planters (Figure 4.11). Zinnia flower seeds were sown in abandoned nests and polyethylene bags to judge the efficiency of nests as seedling sleeves. The criteria for judging were early germination, early two-leaf stage, early nursery raising, flowering, disease incidence, and number of irrigations. Plantlets along with nests are sown on soil at the four-leaf stage. The plastic sleeves of the raised plantlets are removed prior to sowing on soil. Results revealed that the early two-leaf stage was best for nest and plastic sleeves. Nursery raising is initially slow for plantlets sown with nest sleeves compared to those sown without sleeves (Figure 4.10). Plants sown without sleeves were irrigated every alternate day, whereas plants sown along with nests were irrigated every third day. Also, plants sown with nests were

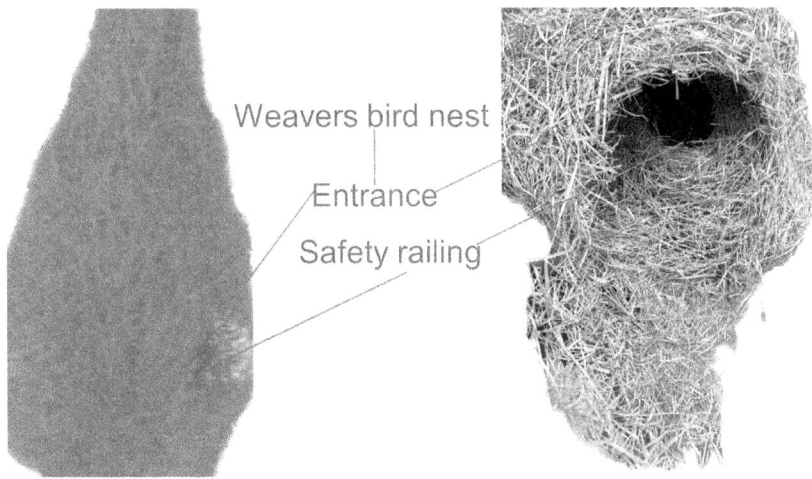

FIGURE 4.9 A close look at the bird-nest, its entrance, safety railing inside to protect the hatchlings.

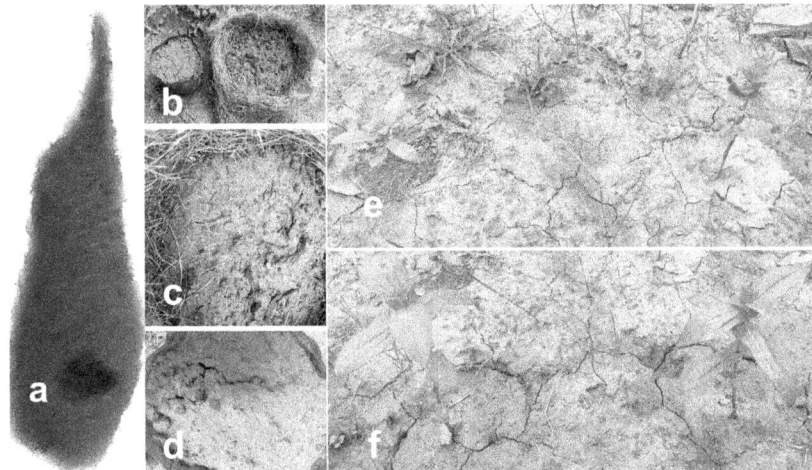

FIGURE 4.10 Use of bird's nest in agro-textiles as seedling sleeve: (a) discarded bird's nest, (b) zinnia flower seeds sown in nest and plastic sleeve, (c-d) two-leaf stage on sleeves, (e) plantlet sown along with nest sleeve, (f) plantlet without sleeve (initially grown on plastic sleeves).

moist for four days by rain irrigation, whereas plants sown without sleeves were moist for only three days. However, the first two-leaf (4th day) and budding (48th day) stages were together for plants sown with and without sleeves. The results indicate that the advantage of using bird nest nursery sleeves is lower irrigation requirements for the plants.

FIGURE 4.11   (a) Four-leaf stage of the plant in nest, (b) plant growing on soil bed along with nest, (c) four-leaf stage of the plant in plastic sleeves, (d) plant removed from plastic sleeve and planted on soil.

FIGURE 4.12   Discarded bird's nests utilized as ecofriendly planters.

4.9   CONCLUSION

In agriculture, nothing is waste. Unutilized discarded agro-residue acts as soil enhancer. Agro-residues are an ideal eco-material, creating many opportunities in the field of technical textiles. The need is to provide avenues to farmers in terms of monetary gains to convince them to stop burning the residue. Going a step further from recycling, the reward for farmers can further the concepts and concerns about saving the environment. Agro-residue's potential in the agrotech, buildtech, meditech, oekotech, and mobiltech industries is well established and sustaining the economy. Agro-residue products provide incentive to farmers and stakeholders. Sustainable products with environmental, social, and cultural values give fruitful results, and there is no other option to achieve eco-efficiency.

REFERENCES

1. Hongu T, Takigami M, Phillips GO. 2005. New Millennium Fibers. Woodhead Publishing Limited, Cambridge, UK. https://doi.org/10.1533/9781845690793.130
2. Byrne C. 2000. Technical textiles market–an overview. In Handbook of Technical Textiles. Woodhead Publishing Limited, Cambridge. 1, 12.
3. Adhia V, Mishra A, Banerjee D, Appadurai AN, Preethan P, Khan Y, de Wagenaar D, Harmsen P, Elbersen B, Van Eupen M, Staritsky I. 2021. Spinning Future Threads: The Potential of Agricultural Residues as Textile Fibre Feedstock. Institute for Sustainable Communities, Research report, Laudes Foundation, Netherlands.
4. Horrocks AR, Anand SC. 2000. Handbook of Technical Textiles. Elsevier, Cambridge, UK.
5. Kozłowski R, Mackiewicz-Talarczyk M (eds). 2012. Handbook of Natural Fibres. Woodhead Publishing, Duxford, UK.
6. Pandey R, Pandit P, Pandey S, Mishra S. 2020. Solutions for sustainable fashion and textile industry. In Pintu Pandit et al. (eds) Recycling from Waste in Fashion and Textiles: A Sustainable & Circular Economic Approach. Scrivener Publishing, Beverly, MA, 33–72. https://doi.org/10.1002/9781119620532.ch3
7. Pandey R, Sinha MK, Dubey A. 2023. Macrophyte and wetland plant fibres. In R Nayak (ed.) Sustainable Fibres for Fashion and Textile Manufacturing. Woodhead Publishing, 109–127. https://doi.org/10.1016/B978-0-12-824052-6.00006-8
8. Sandra N, Alessandro P. 2021. Consumers' preferences, attitudes and willingness to pay for bio-textile in wood fibers. Journal of Retailing and Consumer Services 58:02304. https://doi.org/10.1016/j.jretconser.2020.102304
9. Savoca S, Capillo G, Mancuso M, Faggio C, Panarello G, Crupi R, Bonsignore M, D'Urso L, Compagnini G, Neri F, Fazio E. 2019. Detection of artificial cellulose microfibers in Boops boops from the northern coasts of Sicily (Central Mediterranean). Science of the Total Environment 691:455–465. https://doi.org/10.1016/j.scitotenv.2019.07.148
10. Li Q, Feng Z, Zhang T, Ma C, Shi H. 2020. Microplastics in the commercial seaweed nori. Journal of Hazardous Materials 388:122060. https://doi.org/10.1016/j.jhazmat.2020.122060
11. Mistri M, Sfriso AA, Casoni E, Nicoli M, Vaccaro C, Munari C. 2022. Microplastic accumulation in commercial fish from the Adriatic Sea. Marine Pollution Bulletin 174:113279. https://doi.org/10.1016/j.marpolbul.2021.113279
12. Pandey R, Mishra S, Dubey R. 2023. Luxurious sustainable fibers. In Subramanian Senthilkannan Muthu (ed.) Novel Sustainable Raw Material Alternatives for the Textiles and Fashion Industry. Springer Nature Switzerland, Cham, 57–79. https://doi.org/10.1007/978-3-031-37323-7_4
13. Dris R, Gasperi J, Mirande C, Mandin C, Guerrouache M, Langlois V, Tassin B. 2017. A first overview of textile fibers, including microplastics, in indoor and outdoor environments. Environmental Pollution 221:453–458. https://doi.org/10.1016/j.envpol.2016.12.013
14. Chen G, Feng Q, Wang J. 2020. Mini-review of microplastics in the atmosphere and their risks to humans. Science of the Total Environment 703:135504. https://doi.org/10.1016/j.scitotenv.2019.135504

15. Gasperi J, Wright SL, Dris R, Collard F, Mandin C, Guerrouache M, Langlois V, Kelly FJ, Tassin B. 2018. Microplastics in air: Are we breathing it in? Current Opinion in Environmental Science & Health 1:1–5. https://doi.org/10.1016/j.coesh.2017.10.002

16. Huzaifah M, Sapuan SM, Leman Z, Ishak MR. 2017. Comparative study on chemical composition, physical, tensile, and thermal properties of sugar palm fiber (Arenga pinnata) obtained from different geographical locations. BioResources 12(4):9366–9382.

17. Sharma N, Allardyce B, Rajkhowa R, Adholeya A, Agrawal RA. 2022. Substantial role of agro-textiles in agricultural applications. Front Plant Science 21(13):895740. https://doi.org/10.3389%2Ffpls.2022.895740

18. Razali N, Salit MS, Jawaid M, Ishak MR, Lazim Y. 2015. A study on chemical composition, physical, tensile, morphological, and thermal properties of Roselle fibre: Effect of fibre maturity. BioResources 10(1):1803–1824.

19. Reddy N, Yang Y. 2005. Biofibers from agricultural byproducts for industrial applications. Trends in Biotechnology 23(1):22–27. https://doi.org/10.1016/j.tibtech.2004.11.002

20. Kopania E, Wietecha J, Ciechańska D. 2012. Studies on isolation of cellulose fibres from waste plant biomass. Fibres & Textiles in Eastern Europe 6B(96):167–172.

21. Pappu A, Patil V, Jain S, Mahindrakar A, Haque R, Thakur VK. 2015. Advances in industrial prospective of cellulosic macromolecules enriched banana biofibre resources: A review. International Journal of Biological Macromolecules 79:449–458. https://doi.org/10.1016/j.ijbiomac.2015.05.013

22. Rossi T, Silva PMS, De Moura LF, Araújo MC, Brito JO, Freeman HS. 2017. Waste from eucalyptus wood steaming as a natural dye source for textile fibers. Journal of Cleaner Production 143:303–310. https://doi.org/10.1016/j.jclepro.2016.12.109

23. Abas NF, Abd Rased ANNW. 2016. The thermal performance of manufactured concrete roof tile composite using clay and rice straw fibres on a concrete mixture. Jurnal Teknologi 78(5):451–455.

24. Pandit P, Pandey R, Singha K, Shrivastava S, Gupta V, Jose S. 2020. Pineapple leaf fibre: Cultivation and production. In M Jawaid et al. (eds) Pineapple Leaf Fibers, Green Energy and Technology. Springer Nature, Singapore, 1–20. https://doi.org/10.1007/978-981-15-1416-6_1

25. Ramasamy R, Obi Reddy K, Varada Rajulu A. 2018. Extraction and characterization of calotropis gigantea bast fibers as novel reinforcement for composites materials. Journal of Natural Fibers 15(4):527–538. https://doi.org/10.1080/15440478.2017.1349019

26. Price C, Pedia T. 2020. Application of coir fiber reinforced green composite. *textilevaluechain.in*. August 15. http://textilevaluechain.in/

27. Kale RD, Taye M, Chaudhary B. 2019. Extraction and characterization of cellulose single fiber from native Ethiopian Serte (Dracaena steudneri Egler) plant leaf. Journal of Macromolecular Science Part A 56(9):837–844. https://doi.org/10.1080/10601325.2019.1612252

28. Ma B, Qiao X, Hou X, Yang Y. 2016. Pure keratin membrane and fibers from chicken feather. International Journal of Biological Macromolecules 89:614–621. https://doi.org/10.1016/j.ijbiomac.2016.04.039

29. Mukherjee A, Kabutare YH, Ghosh P. 2020. Dual crosslinked keratin-alginate fibers formed via ionic complexation of amide networks with improved toughness for assembling into braids. Polymer Testing 81:106286. https://doi.org/10.1016/j.polymertesting.2019.106286

30. Sannapapamma KJ, Mariyappanavar S, Sangannavar VV, Jamadar D, Vastrad JV, Byadagi SA. 2020. Development and quality assessment of handmade papers using underutilized agro based natural fibres. Journal of Pharmacognosy and Phytochemistry 9(2):1410–1417.

31. Mendoza RC, Grande JO, Acda MN. 2021. Effect of keratin fibers on setting and hydration characteristics of Portland cement. Journal of Natural Fibers 18(11):1801–1808. https://doi.org/10.1080/15440478.2019.1701604

32. Devi S, Gupta C, Parmar MS, Jat SL, Sisodia N. 2017. Eco-fibers: Product of agri-bio-waste recycling. Journal of Humanities and Social Science 22(9):51–58. https://doi.org/10.9790/0837-2209085158

33. Rajabinejad H, Bucişcanu II, Maier SS. 2019. Current approaches for raw wool waste management and unconventional valorization: A review. Environmental Engineering & Management Journal (EEMJ) 18(7):1439–1456. www.eemj.icpm.tuiasi.ro/; www.eemj.eu

34. Ivanovska A, Veljović S, Dojčinović B, Tadić N, Mihajlovski K, Natić M, Kostić M. 2021. A strategy to revalue a wood waste for simultaneous cadmium removal and wastewater disinfection. Adsorption Science & Technology 1–14. https://doi.org/10.1155/2021/3552300

35. Hassanin AH, Candan Z, Demirkir C, Hamouda T. 2018. Thermal insulation properties of hybrid textile reinforced biocomposites from food packaging waste. Journal of Industrial Textiles 47(6):1024–1037. https://doi.org/10.1177%2F1528083716657820

36. Nam G, Kim J, Song JI. 2019. Mechanical performance of bio-waste-filled carbon fabric/epoxy composites. Polymer Composites 40(S2):E1504–E1511. https://doi.org/10.1002/pc.25063

37. Kamble Z, Behera BK. 2020. Mechanical properties and water absorption characteristics of composites reinforced with cotton fibres recovered from textile waste. Journal of Engineered Fibers and Fabrics 15. https://doi.org/10.1177/1558925020901530

38. Guna V, Ilangovan M, Vighnesh HR, Sreehari BR, Abhijith S, Sachin HE, Mohan CB, Reddy N. 2021. Engineering sustainable waste wool biocomposites with high flame resistance and noise insulation for green building and automotive applications. Journal of Natural Fibers 18(11):1871–1881. https://doi.org/10.1080/15440478.2019.1701610

39. Rathinamoorthy R, Aarthi T, Aksaya Shree CA, Haridharani P, Shruthi V, Vaishnikka RL. 2021. Development and characterization of self-assembled bacterial cellulose nonwoven film. Journal of Natural Fibers 18(11):1857–1870. https://doi.org/10.1080/15440478.2019.1701609

40. Bryson T, Major W, Darrow K. 2001. Assessment of on-site power opportunities in the industrial sector (ORNL/TM-2001/169). Prepared by Onsite Energy Corp. for the Oak Ridge National Laboratory, Oak Ridge, TN.

41. Zheljazkov VD. 2005. Assessment of wool waste and hair waste as soil amendment and nutrient source. Journal of Environmental Quality 34:2310–2317.

42. Górecki RS, Górecki MT. 2010. Utilization of waste wool as substrate amendment in pot cultivation of tomato, sweet pepper, and eggplant. Polish Journal of Environmental Studies 19(5):1083–1087.

43. Agarwal MS. 2013. Application of textile in the agriculture. International Journal of Advance Research in Science and Engineering 2:9–18.

44. Kadam VV, Meena LR, Singh S, Shakyawar DB, Naqvi SMK. 2014. Utilization of coarse wool in agriculture for soil moisture conservation. Indian Journal of Small Ruminants 20(2):83–86.

45. Ghosh M, Biswas D, Sanyal P. 2016. Development of jute braided sapling bag for nursery use. Journal of Natural Fibers 13(2):146–157. https://doi.org/10.1080/15440478.2014.1002147

46. Subramaniam V, Poongodi RG, Veena V. 2009. Agro-textiles: Production, properties and potential. Nonwoven and technical textiles. Indian Textile Journal.

47. Agarwal A, Rastogi M, Singh NB. 2022. Agricultural wastes utilization in water purification. In E Lichtfouse, SS Muthu, A Khadir (eds) Inorganic-Organic Composites for Water and Wastewater Treatment. Environmental Footprints and Eco-design of Products and Processes. Springer, Singapore, vol. 1, 147–168. https://doi.org/10.1007/978-981-16-5916-4_7

48. Khedari J, Noppanun N, Jongjit H, Sarocha C. 2004. New low cost insulating particleboards from mixture of durian peel and coconut coir. Building and Environment 39:59–65.

49. Sampathrajan A, Vijayaraghavan NC, Swaminathan KR. 1992. Mechanical and thermal properties of particle boards made from farm residues. Bioresource Technology 40:249–251.

50. Quintana G, Velasquez J, Betancourt S, Ganan P. 2009. Binderless fiberboard from steam exploded banana bunch. Industrial Crops and Products 29(1):60–66. https://doi.org/10.1016/j.indcrop.2008.04.007

51. Xu J, Sugawara R, Widyorini R, Han G, Kawai S. 2004. Manufacture and properties of low-density binderless particleboard from kenaf core. Journal of Wood Science 50:62–67. https://doi.org/10.1007/s10086-003-0522-1

52. Tangjuank S. 2011. Thermal insulation and physical properties of particleboards from pineapple leaves. International Journal of Physical Sciences 6(19):4528–4532.

53. Ngo TD. 2020. Introduction to composite materials. In TD Ngo (ed.), Composite and Nanocomposite Materials: From Knowledge to Industrial Applications. IntechOpen, London, UK, 2–27.

54. Kalia S, Thakur K, Celli A, Kiechel MA, Schauer CL. 2013. Surface modification of plant fibers using environment friendly methods for their application in polymer composites, textile industry and antimicrobial activities: A review. Journal of Environmental Chemical Engineering 1(3):97–112. https://doi.org/10.1016/j.jece.2013.04.009

55. Zhou XY, Zheng F, Li HG, Lu CL. 2010. An environmental-friendly thermal insulation material from cotton stalk fibers. Energy and Buildings 42:1070–1074.

56. Luamkanchanaphan T, Chotikaprakhan S, Jarusombati S. 2012. A study of physical, mechanical and thermal properties for thermal insulation from narrow-leaved cattail fibers. APCBEE Procedia 1:46–52.

57. Hassan MM, Carr CM. 2021. Biomass-derived porous carbonaceous materials and their composites as adsorbents for cationic and anionic dyes: A review. Chemosphere 265:129087. https://doi.org/10.1016/j.chemosphere.2020.129087

58. Selvi ST, Sunitha R, Ammayappan L, Prakash C. 2023. Impact of chemical treatment on surface modification of agave Americana fibres for composite application–a futuristic approach. Journal of Natural Fibers 20(1):2142726. https://doi.org/10.1080/15440478.2022.2142726

59. Wang B, Yan L, Kasal B. 2022. A review of coir fibre and coir fibre reinforced cement-based composite materials (2000–2021). Journal of Cleaner Production 338:130676. https://doi.org/10.1016/j.jclepro.2022.130676

60. Alves C, Silva AJ, Reis LG, Freitas M, Rodrigues LB, Alves DE. 2010. Ecodesign of automotive components making use of natural jute fiber composites. Journal of Cleaner Production 18(4):313–327. https://doi.org/10.1016/j.jclepro.2009.10.022

61. Dutta AB, Sengupta I. 2015. Sustainable application of agro-waste as substitute for construction materials: A review. IJSART 1(9):30–33.

62. Rusli M, Nanda RS, Dahlan H, Bur M, Okuma M. 2021. Sound absorption characteristics of composite panel made from coconut coir and oil palm empty fruit bunches fibre with polyester. International Journal of Automotive and Mechanical Engineering 18(3):9022–9028. https://doi.org/10.15282/ijame.18.3.2021.14.0691

63. Dunne RK, Desai DA, Heyns PS. 2021. Development of an acoustic material property database and universal airflow resistivity model. Applied Acoustics 173:107730. https://doi.org/10.1016/j.apacoust.2020.107730

64. Yilmaz ND, Banks-Lee P, Powell NB, Michielsen S. 2011. Effects of porosity, fiber size, and layering sequence on sound absorption performance of needle-punched nonwovens. Journal of Applied Polymer Science 121(5):3056–3069. https://doi.org/10.1002/app.33312

65. Berardi U, Iannace G. 2017. Predicting the sound absorption of natural materials: Best-fit inverse laws for the acoustic impedance and the propagation constant. Applied Acoustics 115:131–138. https://doi.org/10.1016/j.apacoust.2016.08.012

66. Bratu M, Ropota I, Vasile O, Dumitrescu O, Muntean M. 2011. Sound-absorbing properties of composite materials reinforced with various wastes. Environmental Engineering and Management Journal 10(8):1047–1051. http://omicron.ch.tuiasi.ro/EEMJ/

67. Yang W, Li Y. 2012. Sound absorption performance of natural fibers and their composites. Science China Technological Sciences 55:2278–2283. https://doi.org/10.1007/s11431-012-4943-1

68. Jayamani E, Hamdan S, Heng SK, Rahman MR. 2014. Sound absorption property of agricultural lignocellulsic residue fiber reinforced polymer matrix composites. Applied Mechanics and Materials 663:464–468. https://doi.org/10.4028/www.scientific.net/AMM.663.464

69. Abdullah AH, Azharia A, Salleh FM. 2015. Sound absorption coefficient of natural fibres hybrid reinforced polyester composites. Jurnal Teknologi 76(9):31–36.

70. Jayamani E, Hamdan S, Rahman MR, Bakri MKB, Kakar A. 2015. An investigation of sound absorption coefficient on sisal fiber poly lactic acid bio-composites. Journal of Applied Polymer Science 132(34). https://doi.org/10.1002/app.42470

71. Jayamani E, Soon KH, Bin Bakri MK, Hamdan S. 2017. Comparative study of sound absorption coefficients of coir/kenaf/sugarcane bagasse fiber reinforced epoxy composites. Key Engineering Materials 730:48–53. https://doi.org/10.4028/www.scientific.net/KEM.730.48

72. Bin Bakri MK, Jayamani E, Soon KH, Hamdan S, Kakar A. 2017. An experimental and simulation studies on sound absorption coefficients of banana fibers and their reinforced composites. Nano Hybrids and Composites 12:9–20. https://doi.org/10.4028/www.scientific.net/NHC.12.9

73. Zhang J, Shen Y, Jiang B, Li Y. 2018. Sound absorption characterization of natural materials and sandwich structure composites. Aerospace 5(3):75. https://doi.org/10.3390/aerospace5030075

74. Chen D, Li J, Ren J. 2010. Study on sound absorption property of ramie fiber reinforced poly (L-lactic acid) composites: Morphology and properties. Composites Part A: Applied Science and Manufacturing 41(8):1012–1018. https://doi.org/10.1016/j.compositesa.2010.04.007

75. Adolphe DC, Schacher L, Dream JY, Khenoussi N. 2016. Overview of some technical textiles. In 16th AUTEX World Textile Conference, June 8–10, Ljubljana.

76. Rezić I. 2013. Cellulosic fibers—Biosorptive materials and indicators of heavy metals pollution. Microchemical Journal 107:63–69. https://doi.org/10.1016/j.microc.2012.07.009

77. Pandey R, Jose S, Sinha MK. 2022. Fiber extraction and characterization from Typha domingensis. Journal of Natural Fibers 19(7):2648–2659. https://doi.org/10.1080/15440478.2020.1821285

78. Dhiman J, Kaith BS. 2020. Fabrication of high performance biodegradable Holarrhena antidysenterica fiber based adsorption devices. Arabian Journal of Chemistry 13(12):8734–8749. https://doi.org/10.1016/j.arabjc.2020.10.004

79. Sheng Z, Gao J, Jin Z, Dai H, Zheng L, Wang B. 2014. Effect of steam explosion on degumming efficiency and physicochemical characteristics of banana fiber. Journal of Applied Polymer Science 131(16). https://doi.org/10.1002/app.40598.

80. Thilagavathi G, Praba Karan C, Das D. 2018. Oil sorption and retention capacities of thermally-bonded hybrid nonwovens prepared from cotton, kapok, milkweed and polypropylene fibers. Journal of Environmental Management 219:340–349. https://doi.org/10.1016/j.jenvman.2018.04.107

81. Rani K, Gomathi T, Vijayalakshmi K, Saranya M, Sudha PN. 2019. Banana fiber Cellulose Nano Crystals grafted with butyl acrylate for heavy metal lead (II) removal. International Journal of Biological Macromolecules 131:461–472. https://doi.org/10.1016/j.ijbiomac.2019.03.064.

82. Futalan CM, Choi AES, Soriano HGO, Cabacungan MKB, Millare JC. 2022. Modification strategies of kapok fiber composites and its application in the adsorption of heavy metal ions and dyes from aqueous solutions: A systematic review. International Journal of Environmental Research and Public Health 19(5):2703.

83. Ilyas RA, Sapuan SM, Ibrahim R, Atikah MSN, Atiqah A, Ansari MNM, Norrrahim MNF. 2019. Production, processes and modification of nanocrystalline cellulose from agro-waste: A review. Nanocrystalline Materials 89–120. https://doi.org/10.5772/intechopen.87001

84. Azril A, Jeng YR, Nugroho A. 2023. Plant-based cellulose fiber as biomaterials for biomedical application: A short review. Journal of Fibers and Polymer Composites 2(1):1–17.

85. Abdul Khalil HPS, Adnan AS, Yahya EB, Olaiya NG, Safrida S, Hossain MS, Balakrishnan V, Gopakumar DA, Abdullah CK, Oyekanmi AA, Pasquini D. 2020. A review on plant cellulose nanofibre-based aerogels for biomedical applications. Polymers 12(8):1759. https://doi.org/10.3390/polym12081759

86. Deepa B, Abraham E, Cherian BM, Bismarck A, Blaker JJ, Pothan LA, Leao AL, De Souza SF, Kottaisamy M. 2011. Structure, morphology and thermal characteristics of banana nano fibers obtained by steam explosion. Bioresource Technology 102(2):1988–1997. https://doi.org/10.1016/j.biortech.2010.09.030

87. Saito T, Nishiyama Y, Putaux J-L, Vignon M, Isogai A. 2006. Homogeneous suspensions of individualized microfibrils from TEMPO-catalyzed oxidation of native cellulose. Biomacromolecules 7(6):1687–1691. https://doi.org/10.1021/bm060154s

88. Meng F, Wang G, Du X, Wang Z, Xu S, Zhang Y. 2019. Extraction and characterization of cellulose nanofibers and nanocrystals from liquefied banana pseudo-stem residue. Composites Part B: Engineering 160:341–347. https://doi.org/10.1016/j.compositesb.2018.08.048

89. Zhou W, Li W, Li J, Zhang Y. 2016. Characterization of cellulose from banana pseudo-stem by polyhydric alcohols liquefaction. Kezaisheng Nengyuan/Renewable Energy Resources 34(2):285–291.

90. Zuluaga R, Putaux JL, Restrepo A, Mondragon I, Ganán P. 2007. Cellulose microfibrils from banana farming residues: Isolation and characterization. Cellulose 14(6):585–592. https://doi.org/10.1007/s10570-007-9118-z

91. Karan CP, Rengasamy RS, Das D. 2011. Oil spill cleanup by structured fibre assembly. Indian Journal of Fiber & Textile Research 36:190–200. http://nopr.niscpr.res.in/handle/123456789/11898

92. Pandey R. 2016. Fiber extraction from dual-purpose flax. Journal of Natural Fibers 13(5):565–577. https://doi.org/10.1080/15440478.2015.1083926

93. Pandey R, Jose S, Basu G, Sinha MK. 2021. Novel methods of degumming and bleaching of Indian flax variety tiara. Journal of Natural Fibers 18(8):1140–1150. https://doi.org/10.1080/15440478.2019.1687067

94. Pandey R, Tiwari N, Dubey A, Jose S, Kambo N, Joshi S, Chauhan VK, Basu, G. 2022. A comparative study of phenotypic variability and physico-mechanical properties of dual-purpose flax fiber varieties in India. Journal of Natural Fibers 19(17):15680–15689. https://doi.org/10.1080/15440478.2022.2133048

95. Eftekhari E, Hasani H, Fashandi H. 2021. Removal of heavy metal ions (Pb2+ and Ni2+) from aqueous solution using nonwovens produced from lignocellulosic milkweed fibers. Journal of Industrial Textiles 51(5):695–713. https://doi.org/10.1016/j.jenvman.2018.04.107

96. Thai QB, Le DK, Do NH, Le PK, Phan-Thien N, Wee CY, Duong HM. 2020. Advanced aerogels from waste tire fibers for oil spill-cleaning applications. Journal of Environmental Chemical Engineering 8(4):104016. https://doi.org/10.1016/j.jece.2020.104016

97. Pandey R, Sinha MK, Dubey A. 2020. Cellulosic fibers from lotus (Nelumbo nucifera) peduncles. Journal of Natural Fibers 17(2):298–309. https://doi.org/10.108 0/15440478.2018.1492486

98. Pandey R, Dubey A, Sinha MK. 2023. Lotus fibre drawing and characterization. In R Nayak (ed.) Sustainable Fibres for Fashion and Textile Manufacturing. Woodhead Publishing, 95–108. https://doi.org/10.1016/B978-0-12-824052-6.00001-9

99. Hassanzadeh S, Hasani H. 2017. A review on milkweed fiber properties as a high-potential raw material in textile applications. Journal of Industrial Textiles 46(6):1412–1436. https://doi.org/10.1177/1528083715620398

100. Madurwar MV, Ralegaonkar RV, Mandavgane SA. 2013. Application of agro-waste for sustainable construction materials: A review. Construction and Building Materials 38:872–878. https://doi.org/10.1016/j.conbuildmat.2012.09.011

101. Ishak MR, Sapuan SM, Leman Z, Rahman MZA, Anwar UMK, Siregar JP. 2013. Sugar palm (Arenga pinnata): Its fibres, polymers and composites. Carbohydrate Polymers 91(2):699–710. https://doi.org/10.1016/j.carbpol.2012.07.073

102. Pandey R, Prasad GK, Dubey A, Arputhraj A, Raja ASM, Sinha MK, Jose S. 2022. Tellicherry bark microfiber: Characterization and processing. Journal of Natural Fibers 19(16):13288–13299. http://dx.doi.org/10.1080/15440478.2022.2089432

103. Manohar K. 2012. Experimental investigation of building thermal insulation using agricultural by-products. British Journal of Applied Science & Technology 2(3):227–239. https://doi.org/10.9734/BJAST/2012/1528

104. Sengupta S, Debnath S, Ghosh P, Mustafa I. 2019. Development of unconventional fabric from banana (Musa Acuminata) fibre for industrial uses. Journal of Natural Fibers 1–13. https://doi.org/10.1080/15440478.2018.1558153

105. Badanayak P, Jose S, Bose G. 2023. Banana pseudostem fiber: A critical review on fiber extraction, characterization, and surface modification. Journal of Natural Fibers 20(1):2168821. https://doi.org/10.1080/15440478.2023.2168821

106. Alavudeen A, Rajini N, Karthikeyan S, Thiruchitrambalam M, Venkateshwaren N. 2015. Mechanical properties of banana/kenaf fiber-reinforced hybrid polyester composites: Effect of woven fabric and random orientation. Materials & Design (1980–2015) 66:246–257. https://doi.org/10.1016/j.matdes.2014.10.067

107. Paul SA, Boudenne A, Ibos L, Candau Y, Joseph K, Thomas S. 2008. Effect of fiber loading and chemical treatments on thermophysical properties of banana fiber/polypropylene commingled composite materials. Composites: Part A 39(9):1582–1588. https://doi.org/10.1016/j.compositesa.2008.06.004

108. Li W, Zhang Y, Li J, Zhou Y, Li R, Zhou W. 2015. Characterization of cellulose from banana pseudo-stem by heterogeneous liquefaction. Carbohydrate Polymers 132:513–519. https://doi.org/10.1016/j.carbpol.2015.06.066

109. Faradilla RF, Lee G, Rawal A, Hutomo T, Stenzel MH, Arcot J. 2016. Nanocellulose characteristics from the inner and outer layer of banana pseudo-stem prepared by TEMPO-mediated oxidation. Cellulose 23(5):3023–3037. https://doi.org/10.1007/s10570-016-1025-8

110. Teli MD, Valia SP. 2013. Acetylation of banana fiber to improve oil absorbency. Carbohydrate Polymers 92(1):328–333. https://doi.org/10.1016/j.carbpol.2012.09.019

111. Cadena Ch EM, Vélez RJM, Santa JF, Otálvaro GV. 2017. Natural fibers from plantain pseudostem (Musa paradisiaca) for use in fiber-reinforced composites. Journal of Natural Fibers 14(5):678–690. https://doi.org/10.1080/15440478.2016.1266295

112. Subagyo A, Chafidz A. 2018. Banana pseudo-stem fiber: Preparation, characteristics, and applications. In Banana Nutrition-Function and Processing Kinetics. Intech Open. https://doi.org/10.5772/intechopen.82204.

113. Ahmad T, Danish M. 2018. Prospects of banana waste utilization in wastewater treatment: A review. Journal of Environmental Management 206:330–348. https://doi.org/10.1016/j.jenvman.2017.10.061

114. Chaitra K, Vinny RT, Sivaraman P, Reddy N, Hu C, Venkatesh K, Vivek CS, Nagaraju N, Kathyayini N. 2017. KOH activated carbon derived from biomass-banana fibers as an efficient negative electrode in high performance asymmetric supercapacitor. Journal of Energy Chemistry 26(1):56–62. https://doi.org/10.1016/j.jechem.2016.07.003

115. Kiruthika AV, Priyadarzini TR, Veluraja K. 2012. Preparation, properties and application of tamarind seed gum reinforced banana fibre composite materials. Fibers and Polymer 13(1):51–56. https://doi.org/10.1007/s12221-012-0051-x

116. Manimaran A, Vatsala TM. 2007. Biobleaching of banana fibre pulp using Bacillus subtilis C O1 xylanase produced from wheat bran under solid-state cultivation. Journal of Industrial Microbiology and Biotechnology 34(11):745–749. https://doi.org/10.1007/s10295-007-0248-y

117. Salamun N, Triwahyono S, Jalil AA, Majid ZA, Ghazali Z, Othman NAF, Prasetyoko D. 2016. Surface modification of banana stem fibers via radiation induced grafting of poly (methacrylic acid) as an effective cation exchanger for Hg (II). RSC Advances 6(41):34411–34421. https://doi.org/10.1039/c6ra03741k

118. Tran LQN, Minh TN, Fuentes CA, Chi TT, Van Vuure AW, Verpoest I. 2015. Investigation of microstructure and tensile properties of porous natural coir fibre for use in composite materials. Industrial Crops and Products 65:437–445. https://doi.org/10.1016/j.indcrop.2014.10.064

119. Yan Y. 2016. Developments in fibers for technical nonwovens. Advances in Technical Nonwovens 19–96. https://doi.org/10.1016/b978-0-08-100575-0.00002-4

120. Rusli M, Irsyad M, Dahlan H, Bur M. 2019. Sound absorption characteristics of the natural fibrous material from coconut coir, oil palm fruit bunches, and pineapple leaf. IOP Conference Series: Materials Science and Engineering 602(1):012067. https://doi.org/10.1088/1757-899X/602/1/012067

121. Parihar CM, Jat SL, Singh AK, Kumar RS, Hooda KS, Chikkappa GK, Singh DK. 2011. Maize Production Technologies in India. Directorate of Maize Research, Indian Council of Agricultural Research, New Delhi.

122. Pandey HP, Kumar A, Pathak RK, Pandey R, Uma Shanker Tiwari US, Mishra RK, Darshan D, Prakash V, Dubey SK. 2022. Biofortification effect of vermicompost, zinc and iron on yield and quality of hybrid maize (Zea mays L.). Agricultural Mechanization in India 53(4):7195–7212.

123. Singh G, Jose S, Kour D, Soun B. 2020. Extraction and characterisation of corn leaf fiber. Journal of Natural Fiber 1–11. https://doi.org/10.1080/15440478.2020.1787914

124. Hossain MM. 2018. Waste management in spinning mills. IOSR-JESTFT 12:1–10.
125. Merli R, Preziosi M, Acampora A, Lucchetti MC, Petrucci E. 2020. Recycled fibers in reinforced concrete: A systematic literature review. Journal of Cleaner Production 248:119207. https://doi.org/10.1016/j.jclepro.2019.119207
126. Reddy N, Yang Y. 2009. Properties and potential applications of natural cellulose fibers from the bark of cotton stalks. Bioresource Technology 100(14):3563–3569. https://doi.org/10.1016/j.biortech.2009.02.047
127. Aspiras FF, Manalo JRI. 1995. Utilization of textile waste cuttings as building material. Journal of Materials Processing Technology 48(1–4):379–384. https://doi.org/10.1016/0924-0136(94)01672-N
128. Jeihanipour A, Karimi K, Niklasson C, Taherzadeh MJ. 2010. A novel process for ethanol or biogas production from cellulose in blended-fibers waste textiles. Waste Management 30(12):2504–2509. https://doi.org/10.1016/j.wasman.2010.06.026
129. Bediako JK, Wei W, Yun YS. 2016. Conversion of waste textile cellulose fibers into heavy metal adsorbents. Journal of Industrial and Engineering Chemistry 43:61–68. https://doi.org/10.1016/j.jiec.2016.07.048
130. Ansell MP, Mwaikambo LY. 2009. The structure of cotton and other plant fibres. In Handbook of Textile Fibre Structure. Woodhead Publishing, vol. 2, 62–94. https://doi.org/10.1533/9781845697310.1.62
131. Mahmoud MA. 2020. Oil spill cleanup by raw flax fiber: Modification effect, sorption isotherm, kinetics and thermodynamics. Arabian Journal of Chemistry 13(6):5553–5563. https://doi.org/10.1016/j.arabjc.2020.02.014
132. Liu D, Han G, Huang J, Zhang Y. 2009. Composition and structure study of natural Nelumbo nucifera fiber. Carbohydrate Polymers 75(1):39–43. https://doi.org/10.1016/j.carbpol.2008.06.003
133. Wang Y, Wang Z, Lu Z, Jung de Andrade M, Fang S, Zhang Z, Wu J, Baughman RH. 2021. Humidity-and water-responsive torsional and contractile lotus fiber yarn artificial muscles. ACS Applied Materials & Interfaces 13(5):6642–6649. https://doi.org/10.1021/acsami.0c20456
134. Cheng C, Du Z, Tan L, Lan J, Jiang S, Guo R, Zhao L. 2018. Preparation and visible-light photocatalytic activity of bismuth tungstate/lotus fiber composite membrane. Materials Letters 210:16–19. https://doi.org/10.1016/j.matlet.2017.08.114
135. Gomez-Campos A, Vialle C, Rouilly A, Sablayrolles C, Hamelin L. 2021. Flax fiber for technical textile: A life cycle inventory. Journal of Cleaner Production 281:125177. https://doi.org/10.1016/j.jclepro.2020.125177
136. Kozlowski R. 2009. Green fibres and their potential in diversified applications. www.fao.org/DOCREP/004/Y1873E/y1873eOb.htm
137. Pillin I, Kervoelen A, Bourmaud A, Goimard J, Montrelay N, Baley C. 2011. Could oleaginous flax fibers be used as reinforcement for polymers? Industrial Crops and Products 34(3):1556–1563. https://doi.org/10.1016/j.indcrop.2011.05.016
138. Sharma HSS, Van Sumere CF. 1992. Enzyme treatment of flax. Genetic Engineering & Biotechnology 12:19–23.
139. Akin DE, Foulk JA, Dodd RB. 2002. Influence on flax fibers of components in enzyme retting formulations. Textile Research Journal 72(6):510–514. https://doi.org/10.1177/004051750207200608
140. Kozlowski RM, Mackiewicz-Talarczyk M, Allam AM. 2012. Bast fibres: Flax. In RM Kozlowski and M Mackiewicz-Talarczyk (eds.), Handbook of Natural Fibres, Vol. 1: Types, Properties and Factors Affecting Breeding and Cultivation. Woodhead Publishing, Duxford, UK, 56–113.

141. Surina R, Andrassy M. 2013. Effect of preswelling and ultrasound treatment on the properties of flax fibers cross-linked with polycarboxylic acids. Textile Research Journal 83(1):66–75. https://doi.org/10.1177/0040517512452928

142. Pandey R. 2021. Extraction of flax fibers by using gel retting method (Indian Patent No. 372965).

143. Richely E, Nuez L, Pérez J, Rivard C, Baley C, Bourmaud A, Guessasma S, Beaugrand J. 2022. Influence of defects on the tensile behaviour of flax fibres: Cellulose microfibrils evolution by synchrotron X-ray diffraction and finite element modelling. Composites Part C: Open Access 9:100300. https://doi.org/10.1016/j.jcomc.2022.100300

144. Melelli A, Jamme F, Legland D, Beaugrand J, Bourmaud A. 2020. Microfibril angle of elementary flax fibres investigated with polarised second harmonic generation microscopy. Industrial Crops and Products 156:112847. https://doi.org/10.1016/j.indcrop.2020.112847

145. De Prez J, Van Vuure AW, Ivens J, Aerts G, Van de Voorde I. 2019. Effect of enzymatic treatment of flax on fineness of fibers and mechanical performance of composites. Composites Part A: Applied Science and Manufacturing 123:190–199. https://doi.org/10.1016/j.compositesa.2019.05.007

146. Shaikh AJ, Varadarajan PV, Sawakhande KH, Pan NC, Srinathan B. 1992. Utilisation of dual-purpose linseed stalk for extraction of fibre (flax) and paper making. Bioresource Technology 40:95–99.

147. Husain K, Malik YP, Srivastava RL, Pandey R. 2009. Production technology and industrial uses of dual purpose linseed: An overview. Indian Journal of Agronomy 54(4):374–379.

148. Mishra S, Pandey R, Singh MK. 2016. Development of sanitary napkin by flax carding waste as absorbent core with herbal and antimicrobial efficiency. International Journal of Science Environment and Technology 5:404–411.

149. Herbig C, Maier U. 2011. Flax for oil or fibre? Morphometric analysis of flax seeds and new aspects of flax cultivation in Late Neolithic wetland settlements in southwest Germany. Vegetation History and Archaeobotany 20:527–533. https://doi.org/10.1007/s00334-011-0289-z

150. Fanti G, Malfi P. 2019. Journey of a flax thread. In G Fanti and P Malfi (eds.), The Shroud of Turin: First Century After Christ! Jenny Stanford Publishing Pte. Ltd., Singapore, 169–189.

151. Pandey R. 2016. History of linen in Indian subcontinent. agropedia.ac.in

152. Zafar MF, Siddiqui MA. 2018. Raw natural fiber reinforced polystyrene composites: Effect of fiber size and loading. Materials Today: Proceedings 5(2):5908–5917. https://doi.org/10.1016/j.matpr.2017.12.190

153. Shuvo II, Rahman M, Vahora T, Morrison J, DuCharme S, Choo-Smith LPI. 2020. Producing light-weight bast fibers from canola biomass for technical textiles. Textile Research Journal 90(11–12):1311–1325. https://doi.org/10.1177/0040517519886636

154. Belachew T, Gebino G, Haile A. 2021. Extraction and characterization of indigenous Ethiopian castor oil bast fibre. Cellulose 28:2075–2086. https://doi.org/10.1007/s10570-020-03667-9

155. Faisal N, Kumar D, Layek A, Kumar N. 2021. Mechanical and thermal behaviour of natural fibers as sustainable green materials. International Journal of Mechanical Engineering 6(3):2610–2616.

156. Ramamoorthy SK, Skrifvars M, Persson A. 2015. A review of natural fibers used in biocomposites: Plant, animal and regenerated cellulose fibers. Polymer Reviews 55(1):107–162. https://doi.org/10.1080/15583724.2014.971124

157. Hrabě P, Mizera Č, Herák D, Kabutey A. 2018. Mechanical behaviour of Sugar palm (Arenga pinnata) fibres. Agronomy Research 16(S1):1046–1051. https://doi.org/10.15159/AR.18.046

158. Bachtiar D, Sapuan SM, Zainudin ES, Khalina A, Dahlan KZM. 2010. The tensile properties of single sugar palm (Arenga pinnata) fibre. IOP Conference Series: Materials Science and Engineering 11(1):012012. https://doi.org/10.1088/1757-899X/11/1/012012

159. Ilyas RA, Sapuan SM, Ibrahim R, Abral H, Ishak MR, Zainudin ES, Asrofi M, Atikah MS, Huzaifah MR, Radzi AM, Azammi AM. 2019. Sugar palm (Arenga pinnata (Wurmb.) Merr) cellulosic fibre hierarchy: A comprehensive approach from macro to nano scale. Journal of Materials Research and Technology 8(3):2753–2766. https://doi.org/10.1016/j.jmrt.2019.04.011

160. Mukhopadhyay S, Fangueiro R, Shivankar V. 2009. Variability of tensile properties of fibers from pseudostem of banana plant. Textile Research Journal 79(5):387–393. https://doi.org/10.1177/0040517508090479

161. Draman SFS, Daik R, Latif FA, El-Sheikh SM. 2014. Characterization and thermal decomposition kinetics of kapok (Ceiba pentandra L.)–based cellulose. BioResources 9(1):8–23. https://doi.org/10.15376/BIORES.9.1.8-23

162. Hazarika D, Gogoi N, Jose S, Das R, Basu S. 2017. Exploration of future prospects of Indian pineapple leaf, an agro waste for textile application. Journal of Cleaner Production 141:580–586. https://doi.org/10.1016/j.jclepro.2016.09.092

163. Lee CH, Khalina A, Lee S, Liu M. 2020. A comprehensive review on bast fibre retting process for optimal performance in fibre-reinforced polymer composites. Advances in Materials Science and Engineering 1–27. https://doi.org/10.1155/2020/6074063

164. Methacanon P, Weerawatsophon U, Sumransin N, Prahsarn C, Bergado DT. 2010. Properties and potential application of the selected natural fibers as limited life geotextiles. Carbohydrate Polymers 82(4):1090–1096. https://doi.org/10.1016/j.carbpol.2010.06.036

165. Saha P, Manna S, Chowdhury SR, Sen R, Roy D, Adhikari B. 2010. Enhancement of tensile strength of lignocellulosic jute fibers by alkali-steam treatment. Bioresource Technology 101(9):3182–3187. https://doi.org/10.1016/j.biortech.2009.12.010

166. Karthik T, Murugan R. 2013. Characterization and analysis of ligno-cellulosic seed fiber from Pergularia daemia plant for textile applications. Fibers and Polymers 14(3):465–472. https://doi.org/10.1007/s12221-013-0465-0

167. Pan Z, Sun D, Sun J, Zhou Z, Jia Y, Pang B, Ma Z, Du X. 2010. Effects of fiber wax and cellulose content on colored cotton fiber quality. Euphytica 173(2):141–149. https://doi.org/10.1007/s10681-010-0124-0

168. Mohanty AK, Misra MA, Hinrichsen GI. 2000. Biofibres, biodegradable polymers and biocomposites: An overview. Macromolecular Materials and Engineering 276(1):1–24. https://doi.org/10.1002/(SICI)1439-2054(20000301)276:1%3C1::AID-MAME1%3E3.0.CO;2-W

169. Ramakrishna G, Sundararajan T. 2005. Studies on the durability of natural fibres and the effect of corroded fibres on the strength of mortar. Cement and Concrete Composites 27(5):575–582. https://doi.org/10.1016/j.cemconcomp.2004.09.008

170. Oldham DJ, Egan CA, Cookson RD. 2011. Sustainable acoustic absorbers from the biomass. Applied Acoustics 72(6):350–363. https://doi.org/10.1016/j.apacoust.2010.12.009

171. Johar N, Ahmad I, Dufresne A. 2012. Extraction, preparation and characterization of cellulose fibres and nanocrystals from rice husk. Industrial Crops and Products 37(1):93–99. https://doi.org/10.1016/j.indcrop.2011.12.016

172. El-Kassas AM, Mourad AI. 2013. Novel fibers preparation technique for manufacturing of rice straw based fiberboards and their characterization. Materials & Design 50:757–765. http://dx.doi.org/10.1016/j.matdes.2013.03.057

173. Sfiligoj-Smole M, Hribernik S, Stana-Kleinschek K, Kreže T. 2013. Plant fibres for textile and technical applications. Advances in Agrophysical Research 369–398. https://doi.org/10.5772/52372

174. Ramdhonee A, Jeetah P. 2017. Production of wrapping paper from banana fibers. Journal of Environmental Chemical Engineering 5(5):4298–4306. https://doi.org/10.1016/j.jece.2017.08.011

175. Mateos-Cárdenas A, Van Pelt FN, O'Halloran J, Jansen MA. 2021. Adsorption, uptake and toxicity of micro-and nanoplastics: Effects on terrestrial plants and aquatic macrophytes. Environmental Pollution 284:117183. https://doi.org/10.1016/j.envpol.2021.117183

176. Rubino C, Aracil MB, Liuzzi S, Stefanizzi P, Martellotta F. 2021. Wool waste used as sustainable nonwoven for building applications. Journal of Cleaner Production 278:123905. https://doi.org/10.1016/j.jclepro.2020.123905

177. Loganathan TM, Sultan MTH, Ahsan Q, Jawaid M, Naveen J, Shah AUM, Talib ARA, Basri AA. 2022. Thermal degradation, visco-elastic and fire-retardant behavior of hybrid Cyrtostachys Renda/kenaf fiber-reinforced MWCNT-modified phenolic composites. Journal of Thermal Analysis and Calorimetry 147:14079–14096. https://doi.org/10.1007/s10973-022-11557-4

5 Eco-Friendly and Sustainable Finishes to Improve Performance of Technical Textiles

5.1 INTRODUCTION

Technical finishes make textiles more serviceable and satisfactory for optimum end use requirements. Except apparel, technical textiles address performance-oriented characteristics of textiles in every possible area from agro-tech (soil) to cryogenic (sky) engineering. The spread of the virus SARS-CoV-2 causing COVID-19 has engulfed the whole world in less than seven months, infecting around 18 million people. The most effective precaution to limit the person-to-person virus spread has been the use of medical masks (N95), surgical masks, and hand-stitched cloth masks. Cloth masks are the least effective among the three but are a sustainable choice for the masses, as surgical masks are melt-blown polypropylene. A shortage of medical and surgical masks also adds to the wide acceptability of cloth masks. In order to transform non-surgical cloth masks into a fabric with enhanced virus deactivating, filtering, and breathing efficiency, suitable technical finishes are imparted. Cloth masks made of fine fibers, compactly woven, with hybrid composition and technical finishes provide efficient filtering of a wide range of particles (10 nm), including bio-aerosols (100–1000 nm), comparable to surgical masks [1]. Efforts are continuously being made to apply finishes on cloth masks to prevent the transmission of viruses and yet maintain proper breathability. Ordinary face masks cause respiratory problems as well as fogging on spectacles and face shields. A electrostatic finish is one of the most effective finishes on cloth masks to capture bio-aerosols (0.1–1 μm). Technical finishes such as superhydrophobicity on face masks form high-density water vapor droplets that roll away from the surface, minimizing the problem of fogging and obstruction in respiration. The insertion of antimicrobial interlining in cloth masks is able to deactivate viruses. Silver nanoparticles, chitosan, plant-based quercetin, eucalyptus oil, or tea tree oil are some of the known bio-based antimicrobial and antiviral finishes for filter masks [1, 2].

A healthy lifestyle and understanding have led to a remarkable increase in healthcare and hygienic products in textiles. The market for textile products with antimicrobial finishes alone is expected to rise above 12.3 billion USD by 2024. UV protection and flame-resistant technical textiles with sustainable features such

DOI: 10.1201/9781003317074-5

as resource saving, recyclability, and biodegradability are valued for their environmental benefits [3]. The quest for multifunctional textile finishes to serve society in every sphere has led to the fabrication of natural superamphiphobic fabrics. Superamphiphobic cotton fabric combined with photocatalysis exhibits excellent dirt repellence, liquid repellence, self-cleaning, ultraslow surface tension, UV shielding, and antifouling performance. Superamphiphobicity is achieved by spraying fluorinated ZnO/HNTs on cotton fabric [4]. A life-saving field weapon for soldiers has been developed utilizing a metal-organic framework with photocatalytic activity to detoxify deadly chemicals used in chemical warfare. The metal-organic framework technical finish built with aluminum powder neutralizes 2-chloroethyl ethyl sulfide (CEES) under visible light radiation [5]. Textile finishes are applied to textiles either through chemical or mechanical processes. Sometimes a combination of strategies is required. The temperature, moisture, and pressure play a key role for getting the desired finish on textiles. Improvement in the performance of textiles involves the application of various finishes and the selection of suitable fiber, thread formation, and weaving structure of the textile materials. Apart from improvement in serviceability, the selection of finishes determines fabric handling and most importantly its end-of-life management. Similarly, microfibers dispersed from fabric during laundering also make their way into the ocean through water streams and are another ecological concern [6]. The presence of finishes and their level of cross-linking on washed-out microfibers affect the adsorption and activity of cellulose-degrading enzymes. Chemical water repellents and durable press finishes decrease the rate of biodegradability of cotton fabric [7]. According to an estimate, 2 million tons of indigo dye (used for dyeing denims) is discharged in effluents every year [8, 9]. The biodegradability of discarded textile material and effluents in the environment is an important issue for a pollution-free environment. Therefore, the application of plant-based finishes that biodegrade along with the substrate without polluting the atmosphere is an important criterion in selecting finishes for textiles. Plant dyes and finishes are sourced from plant parts rich in flavonoids, anthocyanin, carotenoids, and xanthophylls, which imparts antibacterial and ultraviolet protection factors to the fabric [10–12].

Finishes to improve the functionality of textiles are selected based on the end use of the fabric; for example, an umbrella invariably requires water resistance. Textiles used for medicine need the incorporation of an antibacterial finish and high moisture management. Flame-resistant outfits are required by firefighters and industrial wear. The finished material must also preserve the mechanical and comfort characteristics of the original textiles. Simple green technology addressing economical and environmental concerns is required for a cleaner, safe planet [13]. Considering the vast manufacturing and utilization of technical textiles, the use of harmful chemicals to enhance textile properties has far greater implications on the environment. Effluent treatment management also involves a huge cost for large industries. Therefore, large-scale manufacturing of technical textile calls for eco-friendly and sustainable finishes, saving human health and the environment besides cutting the cost of effluent treatment. Sustainable practices like agro-residue dye also impart various technical finishes to the fabric like antibacterial (chamomile extract, chickpea testa) and UV protection (peanut testa) as well as fabric coloring [10–12, 14, 15]. Cellulosic textile processing requires alkali treatment for scouring, degumming, dyeing, and finishing. Rice straw ash, alkaline in nature, used in textile dyeing and finishing is an alternative to alkaline chemicals. Plant dyes and finishes are sourced from plant parts rich in flavonoids, anthocyanin, carotenoids, and xanthophylls, which imparts antibacterial and ultraviolet protection factors to the fabric [16].

Currently, a common concern about plant-based dyes and finishes is their durability on cotton fabric. One of the reasons for the low affinity of natural dyes on cellulosic textile is that cellulosic fibers and plant dye pigments both carry a negative charge. The negative charges of the cellulosic textiles repel the natural dye-containing phenolic hydroxyl group, resulting in the poor dye affinity of cellulosic textiles. The dye affinity of cellulosic textiles may be improved by amine-containing bio-additives such as protein and chitosan [17–20]. Soy protein, containing essential amino acids such as lysine and methionine, has been used as biocolorants, exhibiting enhanced color strength and functionality in the dyed fabric [16]. Red cabbage dye was applied on cotton by incorporating cationic cites by using tannic acid, Rewin Os, Denitex BC, and Sera Fast on cotton fabric. Bioactive red cabbage encapsulated in calcium alginate and potash alum as a fixing agent has been applied on cotton. The process improved the antibacterial properties of the cotton surface, promising for its use as a therapeutic bandage for wound healing [21, 22]. Despite the challenges of cost, fastness, durability, and most importantly availability, consumers are willing to invest a premium price (64 to 128%) on certified biotextiles [23]. Therefore, use of agro-residue parts for dyeing and finishing is advantageous not only for farmers and end users but also for our precious soil, air, and water bodies. A classification of agro-residue is presented in Chapter 4. Efforts to identify sustainable and eco-friendly alternatives to harmful chemical finishes have been made in this chapter. The chemical finishes used in textile processing make it less attractive and have a negative ecological impact. In this regard, enzymes, plants, fungi, bacteria, and plasma treatments are considered environment-friendly technologies for fabric surface modification [24]. This chapter describes technical textile finishes and outlines the biomaterials that impart technical superiority to textile materials.

As discussed in Chapter 4, agro-residues are either left in the field as waste or burned if interfering with tillage operations. Burning is known to deteriorate soil and air quality and thereby human and environmental health. Biomaterials to improve the performance of technical textile materials described in the chapter are enzymes, hydrogels, nanoparticles, dye, mordants, mud, cow dung, and fluorescent proteins. Dye affinity on a textile substrate is improved by mordanting and by dye application methods using plasma and ultrasonic technology [25]. Advanced eco-friendly plasma technology and ultrasonic irradiation, including natural dyes and nanoemulsion used in improving technical finishes, are explained in the chapter. Eco-friendly and sustainable textiles use fewer chemicals in manufacturing and processing. In view of this, finishes sourced from (1) agro-residue, (2) animal proteins, and (1) or (2) in combination with metallic ions are also highlighted.

5.2 AGRO-RESIDUE APPLICABLE IN TEXTILE BIOFINISHES AND DYEING

5.2.1 Bionatural Dye and Mordants

Prior to the 19th century, all the finishes and dyes applied in textiles, buildings, and cosmetics were pristinely sourced from flora, fauna, and minerals [6]. Red colorants from Vincent van Gogh paintings and ancient Andean textiles were investigated by the microspectrofluorimetry method. The results identified the presence of purpurin and eosin in red lake pigments. Red dyes in Andean textiles (Paracas and Nasca textiles) dating from 200 BC to AD 1476 were found to be purpurin, pseudopurpurin, and carminic acid [26]. The durability of ancient dyes can be seen in the brightness of ancient textile artifacts kept in museums, the Ellora cave murals, Kurnool, and the Bhimbetka rock paintings (30,000 years old). Ancient Coptic textiles preserved in a museum in Warsaw were investigated for their coloring compounds using advanced liquid chromatography (LC), ultraviolet-visible (UV–Vis), and mass spectrometric (MS) detection. Flavonoid, anthraquinone, and indigo dye compounds found in 20 ancient samples were luteolin, apigenin, rhamnetin, kaempferol, alizarin, purpurin, xanthopurpurin, monochloroalizarin, and indirubin [27, 28]. Natural dyes and biomordants have taken center stage in textile finishing studies due to their non-irritating behavior towards skin and ecology throughout the production, processing, usage, and disposal chain. Currently global demand for natural dyes is approximately 10,000 tons, which is 1% of synthetic dye usage. A revival of natural dyes in India has been planned by Bureau of Indian Standards (BIS) and Indian Standard Organisation/Textile Committee-38 (ISO/TC-38) since 2015 by forming Working Group 38 (WG-38), which is an active group to finalize global test standards and natural dye certification [29]. Dye mordants are an indispensable part of natural dyeing and used as dye influencers as well as fixers. Pretreatment of fabric with a combination of the natural mordant myrobalan (10%) and metallic salt potash alum (10%) gave sufficient color yield and colorfastness to madder (4%) dyed textiles [29]. Disadvantages of traditional metal mordants are discharge of metal ions in effluent streams, posing environmental danger. Effluents generated by synthetic coloring of textiles have proved hazardous to health and carcinogenic. Countermeasures are to use natural dye sourced from plants [6, 30, 31], animals, minerals [32], and industry effluents [33, 34]. Plant parts such as leaves [35–40], flowers [15, 41, 42], seeds [43], fruits [44, 45], roots [46, 47], and vegetable waste [48] contain dietary fiber, polyphenols, tannins, flavonoids, gallic acid, and carotenoids. Certain fruit-specific components are potential fluorophores used as organic dyes, pigments, emulsions, and fluorescent proteins. The presence of polyphenols in plant extracts is an indication of high UV protection on treated fabric [49]. Additional advantages are their biocompatibility, low cost, and non-toxicity to humans and the environment [50, 51]. Flowers such as marigolds (*Tagetes erecta*) and hibiscus discarded as waste after use in temples and other cultural practices are a source of natural textile dyes for many emerging designers [25, 52, 53]. Marigold petal–dyed textiles are enriched with multifunctional properties such as UPF, moth repellency, antimicrobial properties, fire retardancy, and wrinkle resistance [25]. Peanut testa, an agricultural processing residue largely thrown away as waste, was collected to dye textiles. Dyed textiles exhibited attractiveness, UPF, and antibacterial superiority in cotton dresses (Figure 5.1). Turmeric (*Curcuma longa* L.), belonging to the family Zingiberaceae, is a common herb in Indian cuisine that provides a natural yellow color to food as well as textiles (Figure 5.2). Turmeric rhizome extract is a known

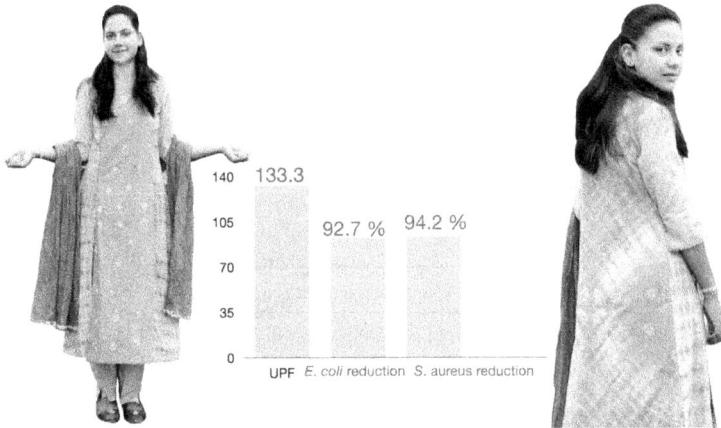

FIGURE 5.1 Peanut dye coloration and its effect on ultraviolet protection factor, bacterial reduction, and aesthetics of cotton fabric.

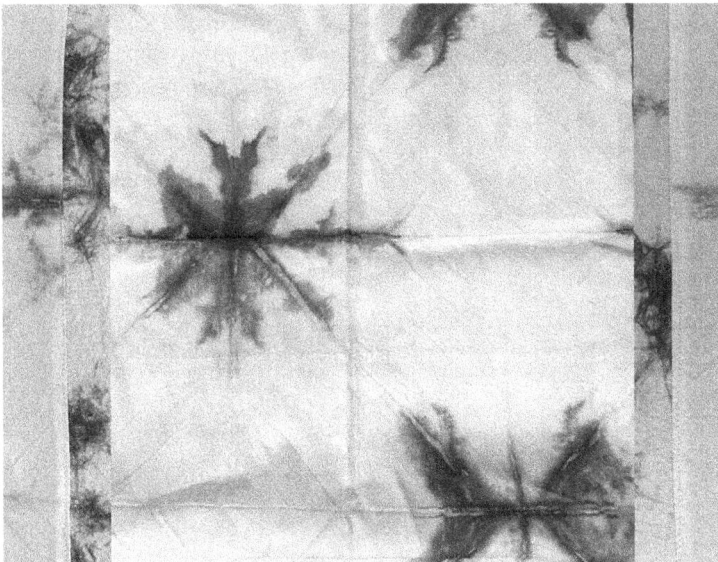

FIGURE 5.2 Turmeric dyeing: cotton tie-dyed with turmeric in combination with reactive dye.

antimicrobial, antifungal, anti-inflammatory, radioprotecting, immunomodulating, antimutagenic, antifibrinogenic, and wound-healing compound [54, 55]. Aqueous extract of turmeric rhizome in the presence of biomordants (acacia, pistachio, pomegranate extract) gave good color strength on fabric [56]. Turmeric- and lac-dyed cotton with a chitosan biomordanting treatment improved fastness, antioxidant, and antibacterial properties. The antioxidant properties of turmeric-dyed cotton with chitosan mordanting reached 61.25% [57, 58]. A combination of pomegranate rind, onion skin, and turmeric root extract had a synergistic effect in dyeing fabric a yellow shade and imparted antioxidant and antimicrobial properties and a UV protection finish [59]. Pomegranate rind contains a higher amount of phenols 307 ± 26 (mg/g extract), flavonoids 150.3 ± 26 (mg/g extract), and tannins 632.6 ± 84.1 (mg/g extract), exhibiting antiradical activity ($11.1 \pm 0.4CI_{50}$ (mgL^{-1}) [60]. Eucalyptus leaf dye on chitosan-pretreated fabric exhibited good color strength, antimicrobial potential, and UV shielding to a great extent [61]. Dye components of plant parts thermally degrade above a certain heating temperature. A differential scanning calorimeter (DSC) is recommended to obtain the optimum temperature for dye pigment stability. Based on DSC thermograph analysis of the color components, the ideal dyeing temperatures for a few plant dyes are: *Butea monosperma* (60–100 °C), madder (60–65 °C), onion (100 °C), and red sandalwood (70–80 °C). A dye lake is formed with the addition of metal salts and natural dye extract. Aluminum-based lake pigments are certified by the Federal Food, Drug and Cosmetic Act (FD&C) and considered safe for human consumption, health, and environment. The thermal stability and colorant concentration of dye lakes are higher than those of plant dye extract solutions, which makes it suitable for technical textile applications. However, dye-lake uptake on textile substrate requires a binder to fix the dye. Bio-based polymers developed as color binders are polylactic acid (PLA) and polyhydroxybuturic acid (PHBA). Other variables that greatly influence the color strength and color fastness are pH, dyeing time, and dye concentration [29, 62]. Table 5.1 presents various natural sources useful for imparting sustainable technical finishes on textile substrates. Synthetic dyeing and finishing additive effluent consists of harmful substances

TABLE 5.1
Natural Sources for Sustainable Technical Finishes on Textile Substrate

Natural Source/ Dye/Oil	Phytochemicals/Heat Resistance	Technical Finish	Reference
Acacia	Catechin, quercetin, tannin	Astringent, UPF	[31]
Aloe vera	Flavonoids, phenols, terpenoids	Antimicrobial, antifungal	[31]
Amaranthus viridis	Flavonoids: quercetin Phenolic: gallic acid	UPF	[31]
Annatto seeds	Bixin, norbixin, lutein	Antifungal, antibacterial, anti-inflammatory	[31]

Natural Source/ Dye/Oil	Phytochemicals/Heat Resistance	Technical Finish	Reference
Banana pseudostem sap	Nitrogen: amino guanidine, mono-ammonium salt, piperidine, carbamic acid, phosphate, phosphite, chloride, tannin	Flame resistant	[31]
Boswellia ovalifoliolata leaf, bark, gum oil	α-amarphene, α-terpineol, β-pinene, β-myrcene, δ-cadinene, caryophyllene oxide, caryophyllene, β-farnesene, α-humulene, ledol, γ-murrolene, (-)-zingiberene	Antibiotic, antibacterial, larvicide, insecticide, antioxidant	[36]
Brassica oleracea L. (kohlrabi) leaves	Catechol, catechin hydrate, chlorogenic, ferulic, p-coumaric, sinapic acids, epigallocatechin, epicatechin 3-O-gallate, syringic acid, gallic acids	Antiradical, antioxidant, antibacterial	[40]
Butea monosperma (flame of the forest)	Butein/60–100 °C	Antimicrobial, antifungal, UV resistant	[29]
Canadian goldenrod (lake pigment)	Quercetin" glucoside, quercitrin, isoquercitrin, rutin, kaempferol, kaempferol-glucoside	High thermal stability (245 °C)	[62]
Chamomile flower	Phenolic, flavonoid content	Antimicrobial, UV protection	[14]
Chickpea testa	Tannin, flavonoid, saponins, phenols, glycosides		[11]
Chitosan	Chitin	Antibacterial (50 washes) Antioxidant (61.25)	[57, 59]
Chitosan + chestnut shell	Chitin + tannin, phenolic acid	UPF	[43]
Coconut shell (green) extract	Tannin, saponin, flavonoid, phenol, terpenoid, glycoside	Self-extinguishing, UV protection, antimicrobial, moth repellent	[31, 48]
Cumin oil	Euganol	Moth repellent	[32]
Eucalyptus oil/leaves	Alpha-pinene, beta-pinene, eucalyptol,8-cineole, benzene, alpha-phellandrene, gamma-terpinene, caffeic acid, nerolidol, linalool, geraniol, limonene, and thymol	Strong antimicrobial, UV protection	[29, 31, 61]
F. artemisiae argyi (FAA) leaves	Eupatilin, Jaceosid	Antimicrobial, UV protection	[16]
Fenugreek seeds	Flavonoid, saponins	UPF, antimicrobial, antifungal	[31]
Henna, walnut	Napthaquinone	Antimicrobial, antifungal	[31]
Hibiscus/roselle flower	Anthocyanin	Antimicrobial	[42]

(Continued)

TABLE 5.1 (Continued)

Natural Source/ Dye/Oil	Phytochemicals/Heat Resistance	Technical Finish	Reference
Indigo ($C_{16}H_{10}N_2O_2$)	Flavonoids, terpenoids, alkaloids, glycosides, indigotine, indirubin, rotenoids	UPF (50), anti-inflammatory, antibacterial, anticonvulsive	[9, 35]
Kohlrabi leaves	Chlorogenic, ferulic, gallic, p-coumaric, sinapic syringic acids	Antibacterial, antioxidant	[40]
Lemongrass	3,7-dimethyl-2,6-octadienal	Antimicrobial, aromatic	[53]
Madder roots	Munjistin, lucidine, rubiadine, purpurin purpuroxanthine, anthraquinones, alizarine/ 60–65 °C	UPF (50), antifungal, mutagenic, antibacterial	[31, 46, 47]
Mangifera indica leaves	Mangiferin, hydrolysable tannin, flavonoids (quercetin), betacyanin, saponin, carotenoids, gallic acid, polyphenol	Antimicrobial, medicinal	[37, 53]
Marigold petals	Carotenoids: lutein, flavonoids: patuletin Polyphenols	UPF, moth repellent, antimicrobial, fire retardancy, wrinkle resistance, antioxidant	[25, 31, 41]
Millettia pinnata + coconut oil + curry leaves	Aromatic groups	Antimicrobial, antifungal (up to ten washes), multifunctional	[38]
Mud	Iron, tannin	Flame resistant, UV protection, antibacterial, hydrophobic	[32]
Neem leaves/seeds	Quercetin, azadirachtin, sallannin, nimbin, alkaloids, flavonoids	Antimicrobial, antifungal, moth repellent	[31, 39]
Onion skin	Flavonoid, quercetin-3,4'-O-diglucoside, quercetin-4'-O-glucoside	UPF, antimicrobial	[50, 62]
Paper industry effluent	Sodium lignosulfonate	UPF (62.13)	[34]
Peanut testa	Saponins, terpenoids, glycosides, flavonoids, phenols, tannin	UPF, antimicrobial	[10, 12]
Pedalium murex Linn + coconut oil + curry leaves	Capric acid, caumarine	Antimicrobial, antifungal (up to ten washes)	[38]
Phyllanthus niruri	Lignan, flavonoids, triterpenes, sterols, alkaloids, essential oils	Mosquito repellant	[30]
Pineapple rind	Carotenoids, gallic acid, polyphenols, flavonoids, mangiferin	Antimicrobial, medicinal	[53]

Natural Source/ Dye/Oil	Phytochemicals/Heat Resistance	Technical Finish	Reference
Pomegranate rind extract	Ellagitannins, flavonoids, polyphenols, gallic acid, ellagic acid, nitrogenous content, amino guanidine, punicalagin, punicalin, tannin	Fire-resistant, UV protection, antimicrobial, antiradical	[31, 44, 60]
Red cabbage	Anthocyanin	Antioxidant, anti-inflammatory, anticancer, antibacterial, antifungal, antiviral	[21, 22]
Solanum nigrum	Flavonoids: quercetin Phenolic: gallic acid	Antibacterial	
Soy protein	Lysine, methionine	Antimicrobial, UV protection	[16]
Tea + polyphenol-melamine-phenylphosphonic acid (TP-MA-PPOA)	Tannin, polyphenol	UV protection (35.2), flame resistance	[48]
Turmeric	Demethoxycurcumin, bisdemethoxycurcumin, diacetyle curcumin, triethyl curcumin, tetrahydrocurcumin, diaroylmethane	Antimicrobial, antifungal, anti-inflammatory, radioprotecting, wound healing	[31, 54, 55, 56]
Tulsi	Caryophyllene, phytol, germacrene antimicrobial compounds	Antimicrobial, medicinal	[31]
Walnut	Napthaquinone	Antimicrobial, antifungal	[31]
Watermelon rind and flesh	Flavonoid, betacyanin, quercetin, β-carotene, tannin	UPF > 50	[45]
Zinc oxide		Antimicrobial, antifungal	[31]

that need to be treated before draining out in water streams [6]. Natural fibers [63, 64], agricultural by-products [65–67], and nanocrystals [68] aid in treatment of textile industry effluents. Toxic textile substance removal by activated biocarbon sourced from seeds [69–71], husks [72–75], leaves [76–79], fruit peels [80–83], and shells [84–90] is presented in Table 5.2.

5.2.2 NANOPARTICLES

Nanoparticle (1–100 nm range) synthesis is done by synthetic chemicals as well as greener methods using plant extracts, sugars, and biodegradable polymers.

TABLE 5.2

Natural Sources for Color Removal from Textile Industrial Effluents

Natural Fibers/ By-Products	Heavy Metal Ions Removed from Aqueous Solution	Adsorption Capacity (mg g^{-1})	References
Almond husk	Nickel	30.769	[72]
Avocado seeds	Chromium	333.33	[70]
Bagasse	Cadmium, nickel, anionic dye	6.194–6.489	[72]
Banana peel	Direct dye, acid dye, indigo, carmine, yellow dye	17.63	[80]
Cactus	Copper, cadmium, iron, dye	98.62	[81]
Cellulose nanocrystal	Cationic dye	243.9	[68]
Coconut shell	Chromium, dichlorodiphenyltrichloroethane (DDT), zinc	Cr-64.49, DDT-14.51	[85, 88, 89]
Coffee	Chromium	87.72	[65]
Corncob	Cadmium	55.2	[72]
Cotton waste	Methylene blue dye	369.48	[66]
Diceriocaryum eriocarpum leaves	Lead	41.49	[77]
Elephant grass	Crystal violet saree	98.42%	[78]
Honeydew peel	Chromium and zinc		[83]
Lime peel	Chromium	100	[82]
Maize tassels	Lead (Pb), vanadium (V), uranium (U), selenium (S), strontium (Sr)	333.3	[76]
Peanut husk	Dye, crystal violet	79.7	[74, 75]
Pistachio shell	Lead, cobalt	81.5%	[86]
Red mud	Cadmium	106.452	[72]
Rice husk	Chromium, bismuth, heavy metals	161.290	[84, 85, 88]
Rubber fig tree leaves	Tetracyclin		[79]
Sapote seeds	Heavy metals, textile dye	59.8	[71]
Sugarcane husk and sawdust	Lead, chromium		[72]
Tea waste	Chromium	94.34	[65]
Teak wood waste	Methylene blue dye	567.52	[67]
Tellicherry bark fiber	Malachite green	99%	[63, 64]
Wood apple shell	Chromium	98.28%	[69]

High surface area and electronic configuration provide the nanoparticles' unique characteristics [91–93]. The textile industry is benefiting immensely from nanoparticles. Nanoparticles are used in textile processing, protective finishes, and production of lightweight and comfortable technical textiles. Nanoparticles impart technical properties on the substrate as well as improving existing properties. The technical potential of nanoparticles embedded in textiles is diverse, including antimicrobial, superhydrophobic, superhydrophilic, water/oil repellent, self-cleaning, flame retardant, electrically conductive, drug-delivering, and UV protection finishes on fabric [93–97]. The presence of bioactive molecules and phytochemical such as alkaloids, amino acids, tannins, enzymes, phenolics, proteins, polysaccharides, saponins, vitamins, and terpenoids in plant fruits, seeds, and peelings functions as a reducing agent in metal nanoparticle synthesis. Some molecules play the role of modeling agents, while a few act as capping agents, aiding in nanoparticle deglomeration [98]. A large list of plants has been compiled by researchers identifying plant extracts suitable for green synthesis of silver nanoparticles [99]. Nanoparticle finishes are eco-friendly, ensuring reduction in application of harmful chemicals on textiles. Nanomaterial finish coatings on textiles improve properties like softness, durability, breathability, water repellency, fire retardancy, protection from ultraviolet (UV) rays, and anti-microbial properties. The bonding strength between nanoparticles and the textile substrate determines the durability of the finish after repeated wash cycles [44, 100–106]. A natural and synthesized melanin nanoparticle (15 nm) finish produced a fabric with an enhanced UV protection layer and photostability, promising for its utilization in optical material applications [107]. Silver nanoparticle synthesis from plant, bacteria, and fungi plays a significant role in various nanotechnological studies due to their superior physico-chemical properties. A silver nanoparticle finish on textiles provides excellent antibacterial and antimicrobial characteristics. The use of silver in drinking water as a disinfectant has been approved by the World Health Organization (WHO); therefore, silver nanotechnology is prominently incorporated in water purification systems [108]. Nanoparticles are synthesized from plants, microorganisms, algae, viruses, fungi, bacteria, and yeast. Their applications include as antibacterial, antimycotic, and anti-cancer agents; as sensors; and in drug delivery. Polyphenols, flavonoids, and sugars in plant extracts are electron-rich phytochemicals capable of reducing metal ions to nanoparticles through redox reactions [109]. Silver nanoparticles are reported to be biosynthesized from *Acalypa indica* leaf extract (5%). Five g leaf extract coated on cotton fabric was able to release silver nanoparticles (42.13% in 200 h) on the surface, making it bacteria free. Silver nanoparticles imparted and prolonged antibacterial properties of cotton fabric [98]. In another study, 12 ml of *A. indica* leaf extract was treated with 100 ml of $AgNO_3$ (1 mM) at 37 °C in dark static conditions. The formation of silver nanoparticles from *A. indica* leaf extract was accomplished in 30 min. Minimum inhibitory concentration (MIC) is the lowest concentration of nanoparticles at which no bacterial growth is visible. In a silver nanoparticle–treated substrate, the MIC was 10 μg/ml against *E. coli* and *V. cholerae*, along with high conductivity in 24 h (235, 215 μS/cm) of the textile [108]. Silver is a historically known biocidal. However, the use of silver nitrate as an antimicrobial finish on textiles stains the fabric on air and light exposure. The drawbacks of silver metal can be overcome by using silver nanoparticles [110]. Silver nanoparticle synthesis from plant and natural sources was found to be highly toxic for multi-drug resistant pathogens and to have superior bactericidal properties at a concentration of 50 ppm [111].

Silver nanoparticles (15 nm) capped with the silk protein sericin imparted antibacterial properties on silk fabric without staining the surface. The use of nanotechnology in biomimicking lotus leaf's effect on the fabric surface by resisting water and oil has potential in self-cleaning of textiles [112]. The layer-by-layer method of deposition of silver nanoparticles capped with polymethacrylic acid created a biofilm on silk and nylon textiles with a potential of 80% and 50% bacterial reduction, respectively. Silver nanoparticle biofilm also colored the fabric to some extent. The intensity of hue due to the biofilm was less in the nylon fabric than in the silk. The results show silver nanoparticles are also able to impart antimicrobial properties on synthetic textiles. Silver nanoclusters improved the luminescence of silk fibers with a 550-nm emission band. Synthesized fluorescent silk fibers possessed antibacterial properties [111, 113–119]. In a simple method, TiO_2 nanoparticle application imparted multifunctional properties, including self-cleaning, UV protection, and antibacterial activity on the textile surface [120, 121]. Cotton immersed in ZnO nanoparticles exhibited highly durable UV protection and antibacterial properties [122].

Plant parts rich in phenolic compounds, flavonoids, ascorbic acid, terpenoids, alkaloids, amino acid, antioxidants, and tannins are used to synthesize silver nanoparticles. Aqueous extractions of plant parts are able to reduce metal ions into nanoparticles [112, 123, 124]. A sustainable and ecofriendly technology of manufacturing nanoparticles from household kitchen waste and metal solutions has been attempted. The synthesized nanoparticles from gold, silver, and zinc have found use in biomedicine and pharmaceuticals. In this method, metal ions are incubated with vegetable and fruit peels to form a colloidal solution for 24 h. Fruit and vegetable peels from banana, orange, mango, pomegranate, watermelon, and potato reduce the size of the metal ions into nanoparticles [98]. Green synthesis of nanoparticles is simple, cost effective, chemical free, high yielding, and environmentally friendly [125–128]. However, cytotoxicity studies reveal nanoparticles synthesized through the green route are more toxic than those from the non-green method [129]. Common nanotechnology used for textile surface modification includes liquid phase deposition, Langmuir-Blodgett films, the sol-gel technique, physical vapor deposition, chemical vapor deposition, ultrasound, and plasma surface modification [130]. Silver nanoparticle synthesis via the green route involves the use of plant extracts, actinomycetes, algal extracts, bacteria, fungi, and yeast. Silver nanoparticles block the multiplication of several bacteria, including *Bacillus cereus*, *Citrobacter koseri*, *Escherichia coli*, *Klebsiella pneumonia*, *Pseudomonas aeruginosa*, *Staphylococcus aureus*, *Salmonella typhii*, *Vibrio parahaemolyticus*, and the fungus *Candida albicans* by attaching silver ions to the microbial cells' biomolecules (Table 5.3). Nanoparticles, smaller in size, easily penetrate the cell structure of the microbes and cause rupture of the cell wall. The size of synthesized nanoparticles is the result of temperature, time, method, and solvent used [129].

5.2.3 ENZYMES

Enzymatic surface modification during textile processing not only modifies the fabric handle, texture, and properties but also introduces functional groups on

TABLE 5.3
Green Synthesis of Nanoparticles from Plant Parts

Plant	Method	Application	Nanoparticle Size (nm)	Reference
Acacia leucophloea	1% bark extract + 95 mL of 1 Mm aqueous AgNO$_3$	Biocompatible, antibacterial	17–29	[123] Murugan et al. 2014
Acalypha indica	Leaf extract + AgNO$_3$ (1 mM)	Pathogen inhibition	20–30	[98, 108]
Aloe vera	10 ml plant extract + 20 mg AgNO$_3$	Strong antibacterial	5–50	[124]
Azadirachta indica	Leaf extract—5 ml AgNO$_3$ (1 mM)—5 ml	Antimicrobial	20–34	[96, 129]
Boerhaavia diffusa	Leaf extract—8 mL AgNO3 (1 mM)—80 ml	Antibacterial	25	[126, 127]
Boswellia ovalifoliolata	Fine powder of bark AgNO$_3$—1 mM	Antibacterial, antifungal	30–40	[113]
Cleistanthus collinus	Leaf extract—1 ml AgNO$_3$ (1 mM)—9 ml	Radical scavenging capacity	20–40	[105, 129]
Cochlospermum religiosum	Bark powder—5 g AgNO$_3$—1 mM	Antibacterial, antifungal	45	[117]
Chitosan+ montmorillonite	Chitosan (100 mL, 0.5 wt%)+ acetic acid+ AgNO$_3$ (0.5–5 g Ag/100 g)	Bacterial resistance	5.42–9.84	[114]
Clitoria ternatea	Leaf extract—5 mL AgNO$_3$ (0.1M)—45 mL	Bactericidal, wound healing, medicine, water purification	20	[126]
Dryopteris crassirhizoma	Rhizome extract	Antimicrobial	5–60	[119, 129]
Garcinia mangostana	Leaf extract: AgNO$_3$ (1 mM) = 1:19	Bactericidal	35	[126]
Gelidiella acerosa	Red seed extract—25 g AgNO$_3$ (1 mM)—90 mL	Antifungal, antibacterial	23	[106, 115]
Leptadenia reticulata	Leaf extract—20 g AgNO$_3$—21.2 g	Antibacterial	50–70	[102]
Millettia pinnata	Flower extract	Cytotoxic, anti-cholinesterase, antibacterial	16–38	[92]
Ocimum sanctum	Leaf extract—5 mL AgNO$_3$ (10^{-3} M)—100ml	Antibacterial and insecticidal	16.87	[94, 101]

(Continued)

TABLE 5.3 (Continued)

Plant	Method	Application	Nanoparticle Size (nm)	Reference
Papaya	Fruit extract—10 ml AgNO₃ (1 mM)— 90 ml	Bactericidal	15	[111]
Pongamia pinnata	Seed extract+ AgNO₃	Antibacterial	16.4	[118]
Shorea tumbuggaia	Fine powder of bark AgNO₃—1 mM	Antibacterial, antifungal	40	[113]
Solanum nigrum	Leaf extract—5 mL AgNO₃ (0.1M)— 45 mL	Wound healing, medicine, water purification	27	[126]
Svensonia hyderobadensi	Fine powder of leaf AgNO₃—1 mM	Antibacterial, antifungal	45	[91, 113]
Syzygium alternifolium	Bark extract—5 mL AgNO₃ (1 mM)—50 mL	Antibacterial	4–48	[103]
Thalictrum foliolosum	Root/leaf extract	Antifungal, biomedical, bioengineered	15–30/18.27	[128]
Vitex negundo	Leaf extract—5 mL AgNO₃ (10⁻³ M)—100 ml	Antibacterial and insecticidal, dye sensitized solar cells	5–47	[94]

the fabric surface [131]. Enzymatic textile finishes and processing are considered environmentally friendly besides being advantageous in terms of resource and energy saving. Ecofriendly processing makes the textiles suitable for further application of technical finishes. Various enzymes are widely used in ecofriendly wet processing of textiles, like transglutaminase treatment on wool, protease for dissolving silk gum, and pectinase for accelerating the retting of bast fibers [132–134]. An eco-friendly, cleaner method developed for degumming or dissolving sericin on silk fibers is by using ultrasonic technology. Enzymes used to degum silk were alcalase, savinase, and their combination. The tensile strength and elongation properties of the silk fabric improved after enzymatic degumming [135]. Glycerol is suggested for dyeing as a replacement for alcohol for environmental considerations and cost effectiveness [136]. Cellulase is useful for removal of noncellulosic impurities from textile surfaces. Cellulase enzyme (2%) application for 60 min at 55 °C on lotus fabric improved the wettability, surface quality, and color absorption of the fabric [38]. Lipases or esterases improve fabric absorbency [93]. Urease, used for biomedical purposes, is the first enzyme that has been crystallized. In immobilized form, urease helps in blood detoxification in kidney machines [137]. The bleaching efficacy of wool fiber is improved by the protease enzyme [31]. Application of the protease enzyme on woolen textiles prior to natural dyeing enhances mechanical properties, water absorbency, dye affinity, and smoothness [138]. Enzymes are also widely used in treatment of textile effluents using biotechnology [133]. Laccase, known as a ligninolytic enzyme, catalyzes dyes and aromatic compounds, accelerating color removal from textile industrial effluents [139].

5.2.4   MUD AND COW DUNG

Silk processing in China using tannin followed by mud smearing has been part of UNESCO's intangible cultural heritage since 2011. Silk soaked in tannin-rich dye is smeared with iron-rich mud. The application of mud on silk is followed by drying the fabric under the sun. The process is repeated ten times. The reaction of tannin and the iron in the mud colors the silk in darker hues with flame resistance and antibacterial properties. The mud-coated side of the silk exhibited strong hydrophobic characteristics [32, 140]. Similarly, cow dung as a green filler has been proposed in composite coating [141]. Cow dung alone and in combination with known mosquito-repellent substances is able to control mosquito infestation without any side effects [142]. Agro-residue such as fibers, seed hulls, and husks are alternatives to synthetic fibrous wool and foams in fabricating insulation boards. Synthetic insulation materials are harmful to skin, lungs, and the environment. Agro-residue ash such as wheat straw ash replaces synthetic fillers in building construction, resulting in lightweight material. The utilization of agro-residue in buildtech is presented in Chapter 4.

5.2.5   HYDROGELS

Hydrogel is a biomaterial composed of nanofibrils useful in medical, industrial, and agro textiles for improved performance. A curcumin-based hydrogelator exhibited promising results in cancer therapy in mice. Hydrogel released 1.9 g/mL of curcumin during an experimental period of 24 h. Hydrogel ensured long-term sustained delivery of curcumin on the substrate and was found to be useful in topical treatment of cancer [143]. Hydrogel from silk protein fibroin was prepared by dissolving 20% fibroin in water and calcium chloride additive. The prepared gel and chamomile extract (5%), mixed under ultrasonic irradiation, was applied on cotton fabric. The results of the antibacterial properties confirmed the antibacterial activity of the hydrogel against gram-positive and gram-negative bacteria [15].

Application of hydrogel in flax retting not only saved time but also improved the quality of extracted fibers in terms of color, tenacity, and consistency in fiber characteristics. Traditional water retting of plant stalks, leaves, and husks is continued for 7 to 30 days [134]. The process results in stench due to long retting duration rendering problem of effluent disposal. Hydrogel in retting water obviates this problem by reducing the retting time, and the effluent generated is safe to drain in crop fields [144].

5.3   ECOFRIENDLY TECHNICAL TEXTILE FINISHES

5.3.1   FLAME AND HEAT PROTECTION FINISHES

Rapid industrialization not only brought a fair share of prosperity to society but also raised safety concerns about workers and environmental issues. Industry workers' clothing needs several features to protect them from occupational hazards such as burning and heat. According to a report by the International Association of Fire and Rescue Services, covering incidents from 1993 to 2014 in 40 countries, representing 15% of the world's inhabitants, there are 5.9 fire injuries and 1.9 fire deaths per 100,000 people. Fire statistics in households concerned the burning of furnishings, upholstery, and nightwear. Academic and industrial studies in flame retardants (FRs) started in 1950. Designing flame retardants for textiles was aimed at lowering the ignition and spread rate of fires [145]. Innovation in fire protection clothing started in 1990, mainly for technical textiles such as curtains; furnishings; and clothing for firefighting, mill, foundry, space, defense, and aviation personnel. Fiber quality and woven structure determine the chemical and physical changes in fabric due to heat and burning. The action of flame retardants applied on a textile substrate either extinguishes the fire or lowers the rate of burning. This is achieved by char formation, decreasing the heat, and depriving the flame of oxygen. Char formation acts as a barrier between the flame and textile substrate, and decreasing heat limits the pyrolysis process. Innovations in flame-retardant materials and process are ongoing, but the golden period of textile flame-retardant studies is regarded as being between 1950 and 1980. Most halogenated flame retardants (decabromodiphenyl or pentabromodiphenyl ethers and poly-chlorinated biphenyls) used initially were banned later due to toxicity to the environment and humans. Currently, the focus is on identifying natural, nontoxic, durable, and ecofriendly flame retardants for textile substrates [31, 145]. Designing flame-protective clothing requires identification of the thermal and burning characteristics of textile fibers. Textile flame causes decomposition, melting, or shrinkage of fibers along with harmful gas and smoke emissions. The fiber content, intensity of ignition, fabric structure, end use, and comfort properties of textiles also play an important role in selecting textile fabrics for application of a flame-retardant finish. Synthetic fibers are thermoplastics having low intensity ignition, slow fire that extinguishes but fiber shrinks and melts making it unstable for use as flame protecting cloth. On melting, a series of degradation products will be yielded by which secondary fire hazards can propagate further fire. Flame resistance of synthetic textiles is achieved by either embedding the FR textiles in a spinning solution or surface treatment [145]. Toxic emission, smoke, thermal stability, air, and water permeability also play an important part in application of fire-resistant finishes on selected fabric. Flame-resistant finishes protect the wearer from fire and prolonged exposure to convection and radiant heat. Non-toxic flame protective finishes enable non-toxic and smoke-free emission with heat contact. Nanocoating on textiles protects the fabric from oil, heat, and flame.

A low limiting oxygen index (LOI) of textiles is increased with the application of flame-retardant chemicals. LOI values below 21 burn easily, whereas those above 21 burn slowly. The fabric is considered flame retardant if the LOI value is above 26–28. Plant-based phosphorus ammonium phytate (APA) is a formaldehyde-free flame retardant that improves the LOI of cotton from 18% to 43.2%. APA forms a covalent bond with cotton, maintaining semi-durable flame retardancy beyond 30 wash cycles with 30.5% LOI [112, 146]. Durable flame retardancy was achieved with the application of nanoparticles made of inorganic metallic salt silica and zinc oxide via the sol-gel method [147]. Alumina and zirconia have also been applied to promote char formation and reduce smoke and burn [148, 149]. After the restriction on halogen-based flame retardants, researchers are working on environmentally friendly non-halogen–based flame retardants. Flame retardants containing phosphorous are also fatal, as they release formaldehyde, a carcinogenic compound [150]. Phytic acid sourced from plant seeds is composed of 28% phosphorous. It is a promising biomacromolecule to provide flame resistance to natural and synthetic textiles. An ammonium alginate, derived from seaweed, and phytic acid combination applied on flax fabric using the layer-by-layer method exhibited excellent flame retardancy [151].

Biomacromolecules containing phosphorus, nitrogen, or sulfur provide a flame-proof substrate (Table 5.4). Whey protein, a bioactive element used in food design, works as a flame-resistant coating on textiles. Its coating on cotton is reported to decrease the burning rate and increase the burning time. Whey protein is reported to act by restricting oxygen diffusion and absorbing combusting heat. $T_{10\%}$ and T_{max1} are the temperatures exhibiting 10% loss and the first thermal degradation stage, respectively. Thermogravimetric data of whey protein coating on cotton shows a decrease in $T_{10\%}$ (°C) and T_{max1} (°C) compared with pristine cotton degradation in nitrogen and air environments. Whey protein and casein decrease the burning rate remarkably and increase final residue (30–80%). Hydrophobins are produced from filamentous fungi. They form hydrophilic and hydrophobic proteins in aqueous media and are able to decrease the burning rate (−13%) on cotton [145]. Phosphorus containing polyphosphates and polyphosphonates grafted on cotton by photochemical atom transfer radical polymerization (photoATRP) modified the fabric, making it highly flame resistant [152]. Organo-phosphorus compound application on cotton enhanced dehydration and char formation with less smoke production. Acid-forming phosphorus compounds are effective in flame retardancy. Phosphorus with nitrogen exhibits a synergistic effect in flame prevention. Finishing formulations in the presence of binders improve the mechanical properties of cotton fabric. Methacryloyloxyethylorthophosphortetraethyldiamidate (P-III) polymer (MPD) was used as a flame retardant on cotton fabric. The treated fabric exhibited smoothness but lower values of air and water permeability in comparison with untreated fabric [150].

Biomacromolecule FR protection effects on textiles are at the pilot scale, where the toxicity is also under appraisal. Limitations on the use of biomacromolecules as textile flame retardants are their limited availability, commercial viability, and wash durability [145].

TABLE 5.4

Effect of Application of Biomacromolecules on Cotton Flameproofing [31, 49, 112, 145, 146, 150, 151]

Biomacromolecules	Constituents	Functional Properties	Burning Time (s)/Rate (mm/s)	LOI (%)
Casein	Phosphorus, phosphate, glutamine	Food ingredients, printing ink, paper making, fibers	75/0.4	18
Coconut shell extract	Cellulose, lignin	Soil quality improvement	72.5	27
Hydrophobins	Cysteine, amino acid residue	Amphipathic, surfactant, pharmaceuticals, biosensors, tissue engineering	104/1.1	
Nucleic acid	Adenine, guanine, cytosine, thymine, deoxyribose	Accelerates char growth	64/1.6	28
Nucleic acid + chitosan	Chitin	Antifungal, medical applications	125	23
Phytic acid + ammonium alginate	Phosphorus, hydroxyl groups	Reduction in flammable volatile and dehydration of fabric		32.2
Whey protein	β-lactoglobulin (β-LG), α-lactalbumin (α-LA), bovine serum albumin (BSA), immunoglobulin (IG), 20% of milk protein	High water absorption, emulsification, good solubility, gelatinization	126/1.0	

Application of banana pseudostem sap on cotton fabric improved the LOI value from 18 to 28–30, resulting in flame-retardant fabric. Cotton fabric was mordanted with tannic acid (5%) and alum (10%) prior to banana pseudostem sap application. The most effective ratio of cotton and non-diluted banana pseudostem sap was found to be 1:10. Soda ash was used to maintain the alkaline pH of the solution. Cotton was dipped in non-diluted banana pseudostem sap for 30 min, followed by quick drying (5 min at 110 °C). The process imparted fire resistance with wash durability and colored the cotton a beautiful khaki hue [31]. Coconut shell extract under alkaline conditions improved the flame retardancy of cotton fabric up to 72.5% (LOI 27). The fabrics also possess strong antibacterial properties and wash durability [49].

5.3.2 ULTRAVIOLET PROTECTION FACTOR

Textiles are one of the most natural barriers to protect human skin from the harmful effects of ultraviolet (UV) rays. The presence of UV blockers on a textile substrate enhances photoprotectivity to act against UV radiation [60]. The ultra-violet radiation range (40 to 400 nm wavelength) is further classified into UV-A (320 to 400 nm), UV-B (290 to 320 nm), and UV-C (200 to 290 nm). UV-A is detrimental to the immunological system, and UV-B is carcinogenic. UV-C is not able to make its way to Earth, as it is absorbed by the ozone layer [25, 49]. The long-term effects of sun exposure are sunburn, skin cancer, and pre-mature ageing. Sun protection is accomplished by sunscreen creams and fabric. Synthetic chemical–based UV absorbers are incorporated in sunscreen creams, lotions, and functional finishes of fabric. Harmful side effects of sunscreen creams and lotions are widely reported. Commonly used chemicals for improving the UPF or sun protection factor (SPF) are phenyl salicylates, oxalic acid, benzo-phenones, benzotriazole, and dianilide derivatives, which are not eco-friendly. Natural dyes are reported to have active molecules that absorb UV rays and block their way through fabric [25, 31]. Other factors that influence UV ray protec-tion are the chemical structure and construction of the fabric. Fabrics with higher cover factor and darker shades have a better ultraviolet protection factor (UPF) than fabrics with low cover and light shades [35]. Aqueous and methanol extracts of *Amaranthus viridis* and *Solanum nigrum* imparted good and excellent UPF on textile substrates. UPF ratings were higher after the application of methanol extract of *Amaranthus viridis* (58.8) and *Solanum nigrum* (>60) plants on fabric than aqueous extract (18.8 and 19.9) [153]. Dye extract from peanut testa [10, 12], chickpea testa [11], babool bark, manjistha roots, Indian madder, indigo [35], pomegranate rind [60], annatto seeds, ratanjot leaves, marigold petals, tree bark [154], and tannin-rich mud [140] have the potential to enhance the UV protection properties of fabric [25]. In the case of peanut testa and madder dye, UPF values are improved with increase in dyeing temperature and dye concentration, respec-tively [10, 155]. Peanut testa dye imparted greater sun protection factor (SPF) in silk fabric (55.73) than in cotton (13.52). The UPF of unmordanted plant dyes is presented in Table 5.5. Sodium lignosulfonate (NLS), an effluent from the paper

TABLE 5.5
Ultraviolet Protection Factor Rating Imparted by Natural Dye on Textiles

Plant Parts/Fabric	Ultraviolet Protection* (UPF)/(Reference)	Shade/Tint Obtained	Reference
Amaranthus viridis/cotton	18.8	White (colorless)	[153]
Andaman satinwood leaf/eri silk	184.42	Greenish-brown	[154]
Burma padauk bark/eri silk	201.94	Brown	[154]
Chickpea testa/cotton–silk–wool	15.58–23.23–78.91	Brown	[11]
Chitosan + chestnut shell–cotton	40	Dark shade	[43]
Cochineal/cotton	36.6	Red	[35]
Indigo leaves/cotton	>50	Blue	[35]
Madder root/cotton–polyester	16.6 [Sarkar]; 50–112	Red	[31, 155]

(Continued)

TABLE 5.5 (Continued)

Plant Parts/Fabric	Ultraviolet Protection* (UPF)/(Reference)	Shade/Tint Obtained	Reference
Marigold petals/wool	>50	Yellow	[25]
Neem bark/eri silk	157.78	Reddish-brown	[154]
Onion peel/cotton	214.19	Yellow	[50]
Peanut testa/cotton–silk–wool	13.52–55.73–65.18	Orange-brown	[10, 12]
Pomegranate rind/cotton	15–24 (good to very good)	Yellow	[60]
Solanum nigrum/cotton	19.9	White (colorless)	[153]

Note: *UPF rating: 15–50 = good to very good, >50 = excellent.

and pulp industry, colored synthetic nylon in darker shades and imparted excellent UV protection (62.13 SPF) on the fabric [34].

5.3.3 ANTIMICROBIAL FINISHES

Antimicrobial finishes on textiles are an important step for protection from skin diseases, odor, and fabric discoloration. Emerging biopolymers such as alginates, chitosan, collagen, hydrogels, hydrocolloids, and superabsorbent polymers have potential in medical applications due to their nontoxic and biocompatible nature [156]. Antimicrobial means either microbicidal or slowing the growth of microbes. This is accomplished by cell wall destruction and by inhibition of the protein and nucleic acid synthesis of the microbes [108, 155]. Antimicrobial activity is due to formation of pits in the microbe cell wall, leading to increased membrane permeability and cell death. Bactericidal action is also attributed to the interaction of silver ions with interior cytoplasm, causing disruption and damage to the cell membrane [108]. Chitosan as a mordant in natural dye exhibited antibacterial properties on fabric [157]. The antimicrobial propensity of chitosan is compromised in cellulosic fabric due to two hydroxyl and one amino group causing poor durability. Quaternization of amino groups modified chitosan, improving its antimicrobial activity. Chitosan grafting on cellulosic by UV curing technology determines the improvement in antimicrobial potential [93]. Chitosan chloride (NMA-HTCC, 1%) in the presence of an alkaline catalyst developed cotton fabric with 100% bacterial reduction against *Staphylococcus aureus* up to 50 washes [57]. A detailed case study of chitosan nanofiber synthesis, properties, and applications is presented in Chapter 6. A silk-based hydrogel and chamomile extract improved the absorption capacity and antibacterial properties of textiles [15]. Natural antimicrobial agents, placed in the advanced antimicrobial category, are set to replace synthetic polymers [93]. Curcumin is widely used as a food additive and textile dye (Figure 5.2). Curcumin is known to have anti-cancerous, antibacterial properties. Curcumin textile dye imparts antibacterial and ultraviolet properties on fabric, suggesting its application in wound dressing [55]. Wet-steamed (15 min) curcumin (8 g/l) and sodium polyacrylate (7.5 g/l) endowed wool fabric with excellent antibacterial activity, color intensity (k/s = 16.15), and fastness [158].

The growth of microbes on textiles is detrimental to the wearer's comfort and the textile itself. Several other negative effects associated with microbes are decolorization of the fabric, loss in mechanical strength, and foul smell. The moisture absorbency of cellulosic fibers makes them more prone to fungal and bacterial infestation. Cotton is largely used in medical textiles for products like bandages, wound dressing gauzes, absorbent napkins, surgical gowns, hospital bed sheets, and pillowcases. Colloidal silver nanoparticles (10 nm) deposited on cotton in 50 ppm concentration exhibit antimicrobial and anti-fungal activity with long-term wash durability. Silver, a known biocidal, is able to inhibit more than 650 different types of microbes. Findings reported silver nanoparticle application post-dyeing should be avoided, as it changes the color of the fabric [110]. Antimicrobial finishes are important for optimum functionality of technical textiles such as meditech and home textiles. The inherent properties of textiles, processing, finishes applied, and climatic conditions affect microbial infestations on fabrics. Microbial infestations on textiles cause pathogen attack and foul odor and make them weak, stained, faded, and unhygienic [31]. Silver (Ag), ZnO, TiO_2, and CuO nanoparticles on textiles act as effective antimicrobial agents, but their reaction with skin and air is under investigation [39]. Silver is an effective disinfectant used for treatment of infectious diseases. A small load of silver ions was found effective in inhibiting bacterial growth on polyester-polyamide by radio-frequency (RF) plasma and vacuum-UV. Silver clusters were found to be deposited on the fabric exhibiting a long-lasting bactericidal effect. Several theories regarding the mechanism of silver ions' antibacterial potential are proposed: (1) interacting with bacterial electron transport; (2) silver ion penetration into bacterial cell and binds bacterial DNA. As a result, replicability of DNA is lost; (3) interference with the cell wall membrane, forming a histidyl complex on the surface, and blocking the dehydrooxygenation process [159]. The use of a silver nanoparticle colloid by using the fungus *Fusarium solani* was attempted. The results exhibited antibacterial efficiency of treated fabric with 54 ppm nanosilver+binder. Cotton fabric showed a reduction in *Staphylococcus aureus* (97%) and *Escherichia coli* (91%) up to 20 washing cycles [160]. Nanoparticles synthesized from aloe vera exhibited fourfold antibacterial propensity compared to commonly known antibiotic medicines [124].

Antimicrobial properties imparted by plant-based dyes are by far the safest option to date. Plant dyes are safe as they are non-toxic and do not harm the skin and environment. Plant dyes containing flavonoids, phenolic compounds, and tannins are an active antimicrobial agent [161]. A list of natural antimicrobial agents that are used on textiles is presented in Table 5.1. Neem oil and seed extract contain quercetin, azadirachtin, sallannin, and nimbin, which is responsible for antimicrobial activity on textiles inhibiting gram-positive and gram-negative bacteria for up to five wash cycles. Neem extract nanoparticles applied on cotton lasted up to 20 washes compared to 10 washes for the neem extract application [31]. Aloe vera's adhesion capacity on cotton is poor. However, the exhaustion method used for applying aloe vera (5%), neem extract (7%), and their combination (10%) together with acetic acid enriched the textile substrate with antifungal (*Aspergillus niger*) and antibacterial (*E. coli, E. aureus*) properties. The wash durability of fabric embedded with a combination of aloe vera and neem was up to 20 wash cycles [93]. Sanitary napkins treated with aloe vera gel exhibited antifungal competence [162]. *Azadirachta indica* leaf extract on lotus fabric caused a 97% reduction in bacterial activity lasting three wash cycles [38]. The application of tulsi and pomegranate extract molecules through the microencapsulation method imparted an antibacterial finish on fabric for 15 washes. Green coconut shell extract applications on fabric not only make the fabric immune to microbes but also protect from ultraviolet rays and flame. Plants obtain nutrients from the soil, including nitrogen, which acts as a catalyst to its flame-retardant property [31]. Essential oils (EOs) such as lemongrass impart microbial inhibition properties [163]. *Milletia pinnata* and *Pedalium murex* Linn in combination with coconut oil and curry leaves make cotton fabric resistant to bacteria and fungi for up to ten washes [39]. Alizarin molecules found in madder dye inhibited bacterial growth (86%) on polyester by damaging the bacterial cell wall [155]. Sustainable green technology of textile finishing using an alginate biopolymer was used to develop antimicrobial finishing. The use of the biopolymer alginate as a green textile finishing using layer-by-layer and sol-gel technology improved the antimicrobial properties of test fabric [164]. Essential oils such as thyme have been reported as antibacterial agents and helpful in inhibition of mold on the textile substrate. Microencapsulation technology was used to apply thyme, patchouli, and peppermint oil on cotton fabric. Gelatin, chitosan, and alginate were used to make an outer shell of microcapsules containing essential oils. The advantages of microencapsulation are controlled release of oil on the substrate and protection from the outside environment [93].

Apart from finish application, durable antimicrobial effects have also been achieved by embedding nanofibers on the textile surface. The process has been done for sportswear with 95% antibacterial activity withstanding up to 35 wash cycles [165].

5.3.4 MOTH AND INSECT PROOFING

Protein fibers like wool, silk, and specialty hair are prone to moth infestation. Dry cleaners and furriers traditionally use commercial chlorinated chemicals like pentachlorobenzene, hexamethylene biguanide, and quaternary ammonium compounds. Concern for eco-friendly and skin-friendly finishing has prompted researchers to look for novel sustainable materials to prevent moths from attacking protein fiber fabrics. In this regard, nanofinishes like nanokeolinite [166] and plant-extracted finishes are a skin-friendly and environmentally sustainable solution. Some notable plant-based moth repellents are neem (leaf, seed, and bark), citronella, clove, lemongrass, peppermint, eucalyptus, and pine oil [31].

Mosquito and insect repellency is required for natural and synthetic textiles alike to protect from diseases like malaria, chikangunia, and dengue. The main chemical-based commercial mosquito repellents for textiles are N, N diethyl 3 benzamide (DEET), pyrethrene, and permythrene. These synthetic chemicals are toxic to mosquitos as well as nature. In contrast, natural plant-extracted moth repellents provide an aromatic fragrance to the finished garment and atmosphere. Biomolecules made from *Vitex negundo* leaves and custard apple extract proved a good mosquito repellant. Tulsi (*Ocimum sanctum*), neem (*Azadirachta indica*), notchi, lemongrass oil (*Cymbopogon lexuosus*), citronella, rui (*Calotropis gigantea*), *Phyllanthus niruri*, cinnamon oil, eucalyptus oil, turmeric, pine oil, garlic, clove oil, peppermint oil, durva grass (*Cynodon dactylon*), and ashoka (*Saraca asoca*) are natural, environment-friendly, aromatic, and non-toxic mosquito repellent chemicals. Mosquitos are not able to survive in contact with clove oil–finished textiles [142].

5.3.5 WATER-RESISTANT FINISHES

The chemical composition and surface geometry of textiles determine its wetting potential. The lotus leaf behavior of picking up surface dirt by liquid droplets and rolling it off the surface has inspired self-cleaning textiles by making them hydrophobic. This is achieved by creating nano-scale–level surface roughness and a low surface energy layer. The surface contact angle is increased by reducing the contact area between water droplets and the surface due to air pockets beneath the liquid. Cellulosic fibers are hydrophilic in nature due to the presence of hydroxyl groups. This hinders the bonding of fibers with the applied hydrophobic matrix. Hydrophobic finishes are therefore required for quality improvement of fabrics [24]. Tannin-rich mud coating, an ancient technology from China, imparts hydrophobicity to silk surfaces [139]. The hydrophobicity of the textiles imparts multifunctional finishes, including antibacterial and UV protection for advanced technical applications. Silver nanoparticles in combination with hexadecyltrimethoxysilane (HDTMS) ($C_{19}H_{42}O_3Si$) produced a superhydrophobic surface on the substrate by reducing surface energy. Treated fibers formed a rough surface, intensifying the thick silver nanoparticle coating on fibers [104]. An ultra-robust superhydrophobic fabric was achieved by immersing the fabric in ZnO nanoparticles and polydimethylesiloxane (PDMS) for 30 min at 110 °C. The treated fabric was resistant to water, corrosive liquids, and knife scathing and provided long-term UV shielding [167].

Multifunctional properties such as superhydrophobicity, biocidal activity, and UV protection (295) were reported with the application of silver nanoparticles on cotton. Superhydrophobicity was observed with a static contact angle more than 150° and water-shedding angle (SHA) of less than 10° [104]. To biomimic the porous hairs of polar bears, scientist prepared a porous textile finished with a superhydrophobic substance. The resultant fabric provided excellent thermal insulation characteristics in air and under water [168]. Cotton textiles exhibited conductive and strong antibacterial properties suitable for biomedical e-textiles

[169]. Superhydrophobicity in silk and cotton was achieved by enzyme etching followed by application of methyltrichlorosilane (CH_3SiCl_3) using the thermo-chemical vapor deposition (70 °) method. The fabric surface developed excellent self-cleaning ability, improved durability, and high water–oil separation efficiency yet retained its intrinsic properties. The process is fluorine free, low cost and environmentally friendly. The enzyme-etched surface exhibited a water contact angle of 156.7° and sliding angle of 8.5°, preserving the physico-mechanical properties, including air permeability. The fabric was resistant to abrasion, with good laundering durability [170]. Graphene oxide, which has amphiphilic properties, exhibited different levels of hydrophobicity by grafting on cotton [171].

5.3.6 ELECTROMAGNETIC INTERFERENCE SHIELDING

Electronic textiles (e-textiles) refers to electrically conductive textile materials. Essential features of e-textiles include light weight, flexibility, comfortability, low cost, and a wearable power source. E-textiles have found potential applications in the field of wearable energy storage, sensors, self-heating fabric, microwave absorption, intelligent textiles, chargers, e-skin, super capacitors, electromagnetic shielding, and global positioning systems (GPSs). Protection from electromagnetic waves is required to protect from life-threatening diseases [13, 172–175]. Conductive material keeps the wearer warm, Nanoparticles made of conductive materials such as silver, copper, gold, zinc, and tin are coated on textiles for enhanced conductivity. Carbon nanotubes, metal oxide, graphite, graphene, and reduced graphene oxide also possess super conductivity [176]. Graphene, a carbon material, has conductivity capability measuring 6000 Scm^{-1}, whereas the conductivity required for protection from electromagnetic waves is 10^{-3} Scm^{-1} [177]. Pyrrole, a low-density (0.9698 @ 20 °C) natural plant-based product applied on cotton, exhibited protection from ultraviolet rays and electromagnetic waves (15.4–62.9 dB for 1 to 3000 MHz frequency) up to 10 wash cycles. The conductive coated cotton was found to be flexible and provided UV protection to the fabric [178, 179].

Conductivity to cotton is imparted by (1) coating of metal nanoparticles, graphene oxide (GO), reduced graphene oxide (rGO), carbon nanotubes (CNTs), poly(3,4-ethylenedioxythiophene) (PEDOT), and metal oxides, which generally possess high conductivity, and (2) cotton fabrication using the nonmetallic conductive polymers dimethyl sulfoxide (DMSO) and poly(3,4-ethylenedioxythiophene: styrene sulfonate) (PEDOT:PSS). Treated cotton has found application in wearable electronics, electrodes, and conducting wires [13].

Carbonization of cotton with oxygen and nitrogen heteroatoms doping led to high electrochemical capability. Prepared lightweight, low-cost, and flexible super capacitors can be used as wearable energy storage devices for charging electronic watches and light-emitting diodes [180]. Graphene oxide reduced by L ascorbic acid, forming reduced-graphene oxide (rGO), was used to dye cotton fabric. The dyed fabric exhibited higher electrical conductivity (2.3 × 10^{-1} S/cm^{-1}), EMI shielding (30–1530 MHz in the X-band as −26 to −35 dB), and improved tensile strength [181]. Graphene-cotton fabric exhibited electrical resistivity with the addition of polymethyl siloxane nanofilament (112.5 kΩ/sq). The treated fabric was superhydrophobic and useful in medical applications, as the fabric surface exhibited self-cleaning, dirt retention, and moisture and corrosion resistance [182]. Metal salts of silver/zinc/copper were used to prepare trimetallic nanoparticles for coating on cotton fabric. Coated cotton exhibited antibacterial capability for 20 wash cycles, UV protection potential, and generous electrical conductivity [183].

Silver nanowire dip-coated on cotton produced conductive surface [184] with antimicrobial potential [185] and lower toxicity [186]. The fabric was proposed for various technical uses such as biosensors, antiseptic wound dressing, and smartech [187]. Chitosan or monochlorotriazinyl-β-cyclodextrin (MCT-β-CD) finishes on textiles improve water uptake capacity by introducing hydroxyl groups in the fabric structure. The presence of moisture enhances the fabric's electrical conductivity and simultaneously its antistatic properties [188]. A bio-based piezoelectric nanogenerator developed from biomaterial (cellulose, silk protein, egg shells, etc.) was integrated into wearable textiles (knee, elbow, wrist, etc.) in the category of smart textiles. In this study, onion skin generated piezoelectricity (~2.8pC/N), providing output voltage (≈18 V), current (≈166 nA), and power density (≈1.7 μW/cm2). The piezoelectric energy conversion potential was ≈61.7%, and it was able to light up to 30 green LEDs. The device was a good sensor for body movements, including coughing, drinking, swallowing, and pulse [189]. Conducting polymer technology has been used on wool, cotton, nylon, and polyester. Techniques adopted for conducting on textiles include exposure of the textile to an oxidizing agent and monomers. Monomers oxidize into radical cations, leading to formation of insoluble oligomers and polymers in solution and on the textile surface. Textile coating is carried out by solution, vapor-phase chemical polymerization, emulsion spraying, brushing, screen printing, and plasma treatment [173].

5.3.7 BIO-PHOTOSENSITIZERS

Photosensitizers with dye from plant parts are able to absorb and convert solar energy to electrical energy. Natural pigments from plant sources are able to be employed in dye-sensitized solar cells (DSSCs), also known as Graetzel cells. Bio-colorants sourced from plants are set to replace metal complex sensitizers. The advantages of natural dye-based sensitizers over metal-complex sensitizers are their low cost, flexibility, biodegradability, availability, and improved performance. Plant pigmentation is due to the interaction of electronic structure of plant pigments and sunlight that result in altering the wavelengths reflected by plant tissues. The transmitted colors of different pigment are the colors perceived by humans and also described as the wavelength of maximum absorbance (λmax) [190]. Flavonoid, carotenoid, anthocyanin, and chlorophyl are the main natural photosensitizers useful as sustainable alternatives for light harvesting in DSSC sensitizers. Flavonoid-rich jatropha leaf dye was used as a DSSC sensitizer for energy production and conservation with (η) 0.12% solar cell conversion efficiency. The photochemical parameters of the jatropha dye–based DSSC sensitizer were λ_{max} (nm) = 400, J_{sc} (mAcm^{-2}) = (0.69), V_{oc} (V) = (0.054), and FF = 0.87 [191].

5.3.8 ILLUMINATING FINISHES

Illuminating finishes are an indispensable part of safety textiles, particularly for traffic police, industry workers, sports activities, cloth counterfeiting, night surveillance, security, and automotive interior features [192]. Traditional illuminating finishes involve the use of chemicals. Recent advances in the field of light absorbing and scattering fluorescence are the use of silk cocoon waste to illuminate textiles. An age-old practice to revive silk sheen is to soak the silk in a metal bath for the purpose of silk weighting. This has prompted scientists to use silk-metal bonding in creating a plasmonic effect in multifunctional smart and wearable textiles. Silk has strong binding properties with metal ions and green chemistry. The combination of fluorescent proteins of silk, metal ions, and green chemicals produced plasmonic color materials using visible light. The advantages of plasmonic silk are its alternative green manufacturing with biocompatibility, scalability, cost effectiveness, and sustainability. Photoluminescent nanomaterials (<10 nm) sourced from agro-residues have the potential to be utilized in biomedical and environmental studies. Pyrolysis, acid treatment, carbonization, nucleation, and oxidation reactions polymerize the discarded fruit and waste peel to carbon dot formation that has found application in energy storage devices, biomedicine, pathogens, heavy metal detection, and purification processes of food and textile effluents [193]. Silver nanoclusters improved the luminescence of silk fibers with the 550 nm emission band. Synthesized fluorescent silk fibers possessed antibacterial properties [116].

5.4 IRRADIATION TECHNOLOGY FOR TECHNICAL FINISHES

5.4.1 PLASMA TECHNOLOGY FOR FINISH APPLICATIONS

Most technical finishes applied on textiles involve the use of treatment techniques such as sol-gel, spray/dip coating, solution immersion, and layer by layer. These processes require non-recyclable organic solvents (cyclic/chlorinated hydrocarbons, alcohols, ketones, tetrahydrofuran, etc.), chemical modifications, and complex techniques (layer by layer) that are hazardous to the environment. Therefore, the need for simple procedures and an eco-friendly and innovative approach in textile finish application has been emphasized [13].

Plasma treatment using nanoparticles on fabric decreases chemical consumption, proving ecologically beneficial [112]. The treatment modifies the natural polymer surface without changing its inherent properties. Low-temperature plasma treatment (power: 1–20 W, pressure: 0.06–0.2 torr, current: 1–6 mA) for 15 s to 10 min improved the wettability of natural fibers. Pretreatment of wool fabric with plasma at low temperatures improved the hydrophilicity of the fabric, resulting in higher dye uptake [194–196]. Plasma is composed of neutrally charged partially ionized gas, ions, and electrons. It is an eco-friendly process widely used to modify the fabric surface without altering its tensile properties. The plasma polymer deposits on the textile surface and acts as a barrier to washing; thus the finish performance properties last longer than those from conventional finishing methods. An experimental fabric treatment with titanium dioxide/silicon dioxide exhibited an ultraviolet protection factor value above 50 but was able to last only 5 wash cycles, whereas hexamethyldisiloxane (HMDSO) plasma polymer–induced UPF could withstand 20 wash cycles. Similarly, a HMDSO plasma polymer contributed to improved long-term antibacterial properties, flame retardancy, thermal stability, and crease recovery of fabric compared to traditional fabric finishing without plasma polymer [31]. Application of plasma and enzyme treatment results in etching on the textile surface. Etching on the cotton textile surface causes removal of non-cellulosic compounds, leading to its hydrophilicity [197].

5.4.2 ULTRASONIC IRRADIATION

Ultrasound irradiation causes the appearance of microbubbles, which oscillate with high pressure. The frequency range of ultrasound is 20–100 kHz (power ultrasound) and 1–10 MHz (diagnostic). The process generates microjets near the textile substrate, resulting in increased heat and diffusion of the solution mix inside the fabric surface. Dye extraction efficacy from natural plant materials is improved (13% to 100%) by using ultrasound in comparison with magnetic stirring. The eco-friendly ultrasonic technique of obtaining useful plant extracts is more effective than traditional heating methods, especially given current global environmental concerns [33]. The method is useful in extracting dye and tannins from plants, achieving good fastness and improved dyeability [40]. Ultrasonic irradiation is the most effective mordanting bath, yielding a softer handle with less weight gains in the resultant fabric. In an experiment on dyeing cotton with a combination of soy protein (3%) and *F. artemisiae argyi* (FAA) leaves, an ultrasonic bath was to be found better than a water or magnetic stirrer bath. A soy protein mordant improved the color strength of dyed cotton, whereas FAA is known to possess antibacterial activity. The resultant dyed cotton possessed dye affinity as well as strong antibacterial potential (100% bactericidal). The ultrasonic method of dye application on cotton produced a soft surface, higher color strength (K/S), and lower increase in weight of the dyed material as compared to water and magnetic stirring baths. The fabric stiffness of the dyed fabric was lower in the ultrasonic method, whereas dyeing in a water bath produced a rough and stiff surface. Thus, ultrasonic is an ideal method of achieving a soft fabric handle and improved ultraviolet protection. The ultrasonic bath diffused the dye and mordants into fiber more effectively. Increased dye intensity and greater coating of soy protein improved the fabric's UV protection and antibacterial properties, respectively. However, strength loss (4–6%) and an increase in fabric extensibility (10%) were also observed in fabric dyed in an ultrasonic bath [16]. The strength loss in an ultrasonic bath (4–6%) using soy protein on cotton is less than the strength loss (19.01%) due to peanut testa protein containing an aqueous dye bath (19.01%) [10].

5.5   CASE STUDIES OF TEXTILE TECHNICAL FINISHES DERIVED FROM PLANT SOURCES

5.5.1   PEANUT TESTA DYE

In a recent study cotton, silk, and wool textiles were dyed using peanut testa (5%) without any chemical intervention. Peanut testa dye was obtained by the aqueous extraction method at 100 °C for 30 min. The extracted dye was used to dye (Figures 5.1 and 5.4b) the selected textiles with a M:L ratio of 1:15 at 90 °C for 30 to 45 min [10]. Lyophilized crude peanut testa powder (Figure 5.4a) (yield 22.8%) was used for printing textiles (Figure 5.4c). The results in Figures 5.5 and 5.6 show that the brightness and color strength in wool and silk textiles is more than in cotton. The reason for the brighter hues in silk and wool protein fibers is the presence of an amine group that bonds well and forms complexation with the negative charges of the phenolic hydroxyl group of natural dye. In contrast, the presence of hydroxyl groups in cotton carries a negative charge that repulses the phenolic hydroxyl group of the natural dye, resulting in poor dye affinity of cellulosic fibers towards natural dye [16]. Chitosan and protein (whey and soy) as mordants are suggested to overcome the poor dye affinity of cellulosic textiles [17–20, 198]. The findings of the research revealed 750 g of peanut testa is required to dye 1000 g of fabric. Peanut testa yield from shelled peanuts was 3.5 to 4.5% [10]. This finding, along with present peanut production data (FAOSTAT), was used to calculate the peanut testa (3% of shelled peanut) dyeing potential (Table 5.6). Keeping in mind the current availability of peanut testa (47.1 million tons/year) in the world (FY 2017), 1.8 million kg of fabric can be colored [12]. Phytochemical analysis (Table 5.7) of peanut testa confirmed the presence of bioactive constituents such as phenols, flavonoids, tannins, saponins, terpenoids, and glycosides (Figure 5.6a, Table 5.8) [198–201]. The ultraviolet-visible spectral range of peanut skin extract was 465–560 nm, with a maxima at 440 nm (Figure 5.6b). The absorbance of anthocyanin falls in this region. Dye uptake (Table 5.9) was highest for silk, followed by wool and cotton fabric. The K/S value of the dyed fabric increased with higher dye concentrations [10]. Aqueous extraction of peanut testa dye was free from any chemical additives and reagents and therefore safe to discharge. Cotton fabric exhibited various shades in combination with peanut testa dye and different natural mordants (Figure 5.3). Post-mordanting gave superior results compared to pre and meta-mordanting in cotton. The physical parameters of peanut testa dyed fabric were altered compared to undyed (Table 5.10). Fabric dyeing lowered the fabric strength in aqueous (strength reduction: cotton 19.01%, silk 35.45%, wool 7.30%) as well as in dyeing methods using microwave energy (strength reduction: cotton 30.12%, silk 62.20%, wool 12.38%) (Table 5.11). Polyphenol, saponin, and tannin present in peanut testa extract incorporated ultraviolet blocking properties on dyed fabric [10]. Abundantly available low-cost peanut testa imparted high ultraviolet protection (Figure 5.1), good wash-fastness (Table 5.9 and 5.10), and antimicrobial superiority (Figure 5.1), making it a choice for formal (Figure 5.1) and casual wear alike [12].

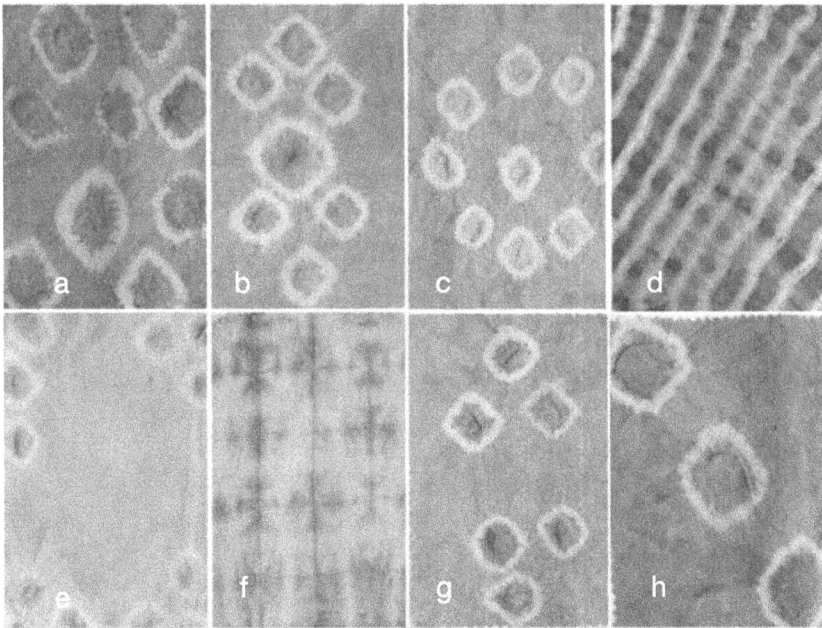

FIGURE 5.3 Peanut-dyed cotton fabric followed by meta-mordanting using (a) myrobalan, (b) banana flower, (c) coffee, (d) alum, (e) tea leaves, (f) chana husk, (g) banana pseudostem sap, (h) *Terminalia bellirica*.

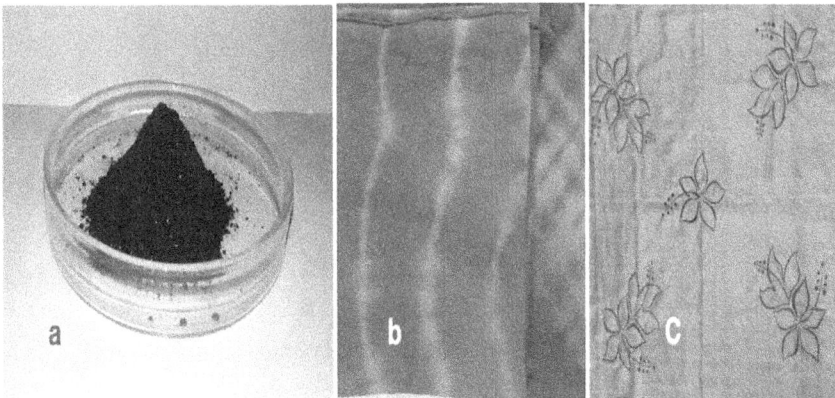

FIGURE 5.4 (a) Lyophilized crude dye, (b) tie-dyed cotton using peanut testa extract, (c) crude peanut testa dye prints on silk.

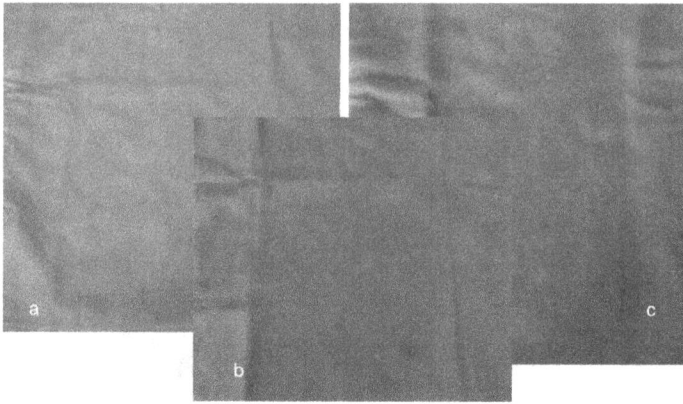

FIGURE 5.5   Peanut-testa dyed textiles: (a) cotton, (b) silk, (c) wool.

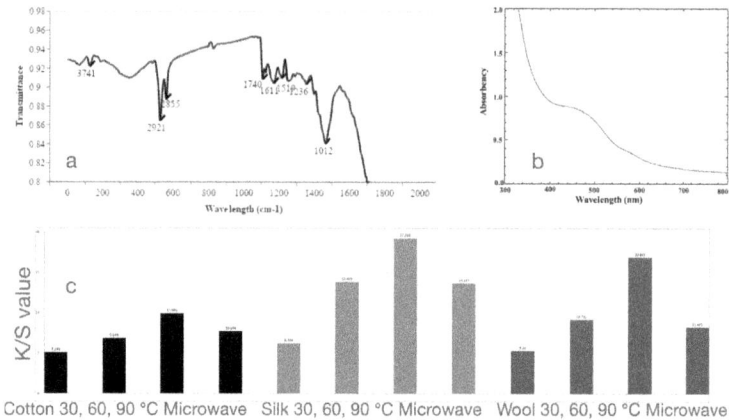

FIGURE 5.6   (a) FTIR graph of peanut testa, (b) dye absorbency, (c) K/S values of dyed textiles at different temperatures.

TABLE 5.6

Peanut Testa Availability (FY 2020) and Dye Economics (FAOSTAT)

Countries (Top 5)	Peanut Production (Shelled) (Thousand Metric Tons)	Peanut Testa Availability (Thousand Metric Tons)	Fabric Dyeing Potential (Thousand Metric Tons)
China	17,500.00	525	700.00
India	6,500.00	195	260.00
Nigeria	3,900.00	117	156.00
United States	2,782.00	83.46	111.28
Sudan	1,800.00	54.00	72.00

TABLE 5.7
Presence of Functional Groups in Peanut Testa [10, 12]

Peak	Functional groups
1012 cm^{-1}	Saponin (ester group)
1236 cm^{-1}	Phenolic (O–H stretching)
3344 cm^{1}	
3740 cm^{-1}	
1519 cm^{-1}	Aromatic ring
1536 cm^{1}	C=C stretching from alkenes
1611 cm^{-1}	
1740 cm^{-1}	Saponin (aliphatic acetoxy group)
3260–3270 cm^{1}	Tannin
1450–60 cm^{1}	Tannin (O-H bending)
2912 cm^{1}	C-H stretching (terpenoids)
2855 cm^{-1}	

TABLE 5.8
Presence of Phytochemicals in Peanut Testa

Bioactive Constituents of Peanut	Identification Method (Manual)	Antibacterial Mechanism	Reference
Flavonoids	Zinc-hydrochloride test (PT extract + zinc dust + con. HCl = red color)	Solubilize extracellular proteins and bind bacterial cell wall	[10, 12, 199]
Glycosides	PT extract + few drops of glacial CH$_3$COOH + 2% FeCl$_3$ + 2 ml of con. H$_2$SO$_4$ = brown ring	Suppress bacterial growth by preventing bacterial RNA production	[10, 12, 202]
Phenolic acid	PT extract + blue litmus paper turns red	Antioxidant ability	[10, 12, 201]
Polyphenols	PT extract + bromine water = white precipitate	Antibacterial ability	[10, 12, 201]
Tannins	2 ml PT extract + 2 ml 2% FeCl$_3$ = blue-black	Inhibit bacterial cell proliferation and inactivate functional cell proteins	[10, 12, 200]

TABLE 5.9

Color Strength and Fastness Properties of Peanut Testa–Dyed Textiles (Unmordanted) at Optimized Temperature and Microwave Energy [10]

Fabric	Dye Absorbance	Dye Uptake (%)	K/S	UVA(315–400 nm)	UVB(290–315 nm)	UPF(290–400 nm)	Rubbing Fastness (Acid/Alkali)	Perspiration Fastness (Acid/Alkali)	Wash Fastness	Light Fastness
Cotton (90 °C)	0.6778	30.48	13.9	10.62	7.64	11.98	4 / 4/5	4/5 / 4	3/5	5/6
Cotton (microwave)	0.7447	20.61	10.8	9.35	6.78	13.52	4 / 4/5	4 / 3/4	3/4	6
Silk (90 °C)	0.4318	64.47	27.0	2.41	1.61	55.73	4 / 4/5	4 / 4	3/4	6
Silk (microwave)	0.4089	67.10	19.2	3.54	2.18	40.13	3/4 / 3/4	3/4 /3/4	4	6
Wool (90 °C)	0.6576	32.89	23.7	3.40	1.15	65.18	4/ 4/5	4/5–4	4	5/6
Wool (microwave)	0.7447	20.61	11.6	3.98	1.2	61.38	4/ 4/5	4/ 3/4	4	6

TABLE 5.10

CIELab, Color Strength, and Fastness Properties of Mordanted Peanut Testa–Dyed Cotton Fabric [12]

Sample Description	L*	a*	b*	K/S	Wash Fastness	Light Fastness	Rubbing Fastness (Dry)	Rubbing Fastness (Wet)
Without mordant	73.7	10.6	16.6	1.1	3	4	5	5
Alum pre-mordant	76.4	6.5	13.5	1.2	3–4	5	5	5
Alum post-mordant	76.6	6.1	13.4	1.5	3–5	5	5	5
$FeSO_4$ pre-mordant	58.9	3.6	14.6	3.8	3–4	5–6	4–5	4
$FeSO_4$ post-mordant	60.2	2.7	15.4	4.1	4	5–6	4–5	4–5
Myrobalan pre-mordant	74.6	1.4	12.9	1.3	3–4	5–6	4–5	5
Myrobalan	73.2	4.3	13.9	1.9	4	5–6	5	5

TABLE 5.11

Fabric Parameters of Peanut Testa–Dyed Textiles [10]

Fabric Parameters	Cotton (Undyed)	Cotton (Dyed)	Silk (Undyed)	Silk (Dyed)	Wool (Undyed)	Wool (Dyed)
EPI	59	63	104	106	47	47
PPI	58	60	96	101	38	41
GSM (g/sqm)	145	155	48	56	189	174
Fabric tenacity (cN/tex)	4.05	3.28	18.84	12.16	3.15	2.92
Wicking height (cm)	5.1	4.7	2.3	1.4	1.7	4.433
Water absorbance (%)	160	65.5	149	124	221	125.8
Water drop (s)	0.496	0.70	6.0	1.3	60.0	7.3

5.5.2 CHICKPEA TESTA DYE

Chickpea testa is also an agricultural processing residue similar to peanut testa. Roasted chickpea and peanut are used to prepare flour and several snacks for human consumption, whereas testa of both grains are discarded. Chickpea testa is reported to be marketed for mixing with cattle feed [11].

Dye preparation method:

Batch I – 100 g chickpea testa + 1 liter water $\xrightarrow{\text{Boil 1 h}}$ Strain extracted dye liquor

Batch II – Filtrate chickpea testa + 1 liter water $\xrightarrow{\text{Boil 1 h}}$ Strain extracted dye liquor

Batch I + II = Dye liquor + extra water to prepare 2 liters of dye solution (5% chickpea dye solution), material: liquor = 1: 15

Tannin present in chickpea testa aqueous extract binds firmly with the textile substrate (covalent bond), giving it a reddish brown (Figure 5.7b–d) fast color

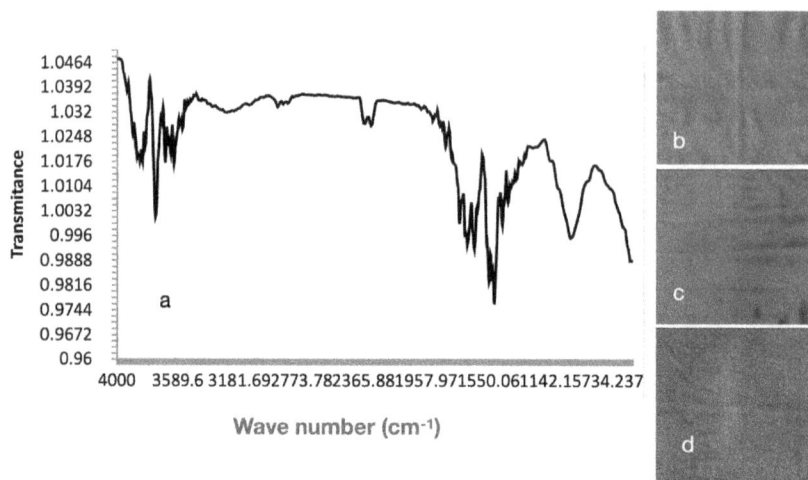

FIGURE 5.7 FTIR graph of chickpea testa dye and the dye color on (a) cotton, (b) silk, (c) wool.

(Table 5.12). A FTIR graph of chickpea dye extract exhibited the presence of functional groups such as tannins, saponins, glycosides, phenols, and flavonoids (Figure 5.7a). The presence of phytochemicals such as polyphenols and tannins also imparted ultraviolet protection (Table 5.12), antimicrobial, and antioxidant characteristics to the dyed textiles. Chickpea testa extract dye on cotton (c) and silk (s) achieved 95.77c – 99.17s% and 97.35c – 98.27s% reduction in *E. coli* and *S. aureus*, respectively. The bacterial reduction imparted by chickpea testa is slightly more than that of peanut testa (Figure 5.1) transmission on cotton [11]. The results thus recommend the use of chickpea and peanut testa dyed textiles for patient bedsheets and clothing in hospitals.

5.5.3 NANOEMULSIONS FROM ONION PEEL

Onion peel, a discarded waste, was collected for study from household and commercial kitchens. To prepare dye (alcohol extraction), the red onion peel is dried in shade and ground to form powder. Twenty-five g onion peel powder and 250 ml ethanol is refluxed in a soxhlet apparatus. The solvent is evaporated using a rotary evaporator, yielding 0.7 g dye extract [50]. An onion peel nanoemulsion was made by mixing the dye extract (dropwise) and the solution of polysorbate 80 (Tween 80) + Sorbitan monooleate (Span 80) and distilled water in a 1:3 ratio. Stirring was continued until a homogenous and stable solution was formed. The onion peel nanoemulsion (Figure 5.8a) dyed the fabric a reddish hue (Figure 5.8b). Functional groups present in the emulsion were quercetin and tannin compounds (Figure 5.8c). The prepared nanoemulsion (5–10%) was used to dye cotton and polyester-cotton blended fabric in a material to liquor ratio of 1:25. A 10.4–11.9% reduction in tensile strength of cotton and 7.5 to 8.6% reduction in

TABLE 5.12

Color Strength, Ultraviolet Protection Factor, and Fastness Properties of Chickpea Testa–Dyed Textile (Unmordanted) at Optimized Temperature (90 °C) [11]

Fabric	L*	a*	b*	K/S	UV-A (315–400 nm)	UV-B (290–315 nm)	UPF (290–400 nm)	Rubbing Fastness (Acid/Alkali)	Perspiration Fastness (Acid/Alkali)	Wash Fastness	Light Fastness (Sunlight)
Cotton	66.3	4.9	11.5	0.88	8.15	6.15	15.58	4 / 3	3/5 / 3	3/4	3/4
Silk	60.2	6.4	13.4	2.52	5.62	3.86	23.23	4 / 3/4	4 / 3/4	3/4	4
Wool	60.4	5.8	13.1	3.43	1.35	0.39	184.66	4 / 3	4/5 / 4	4	4

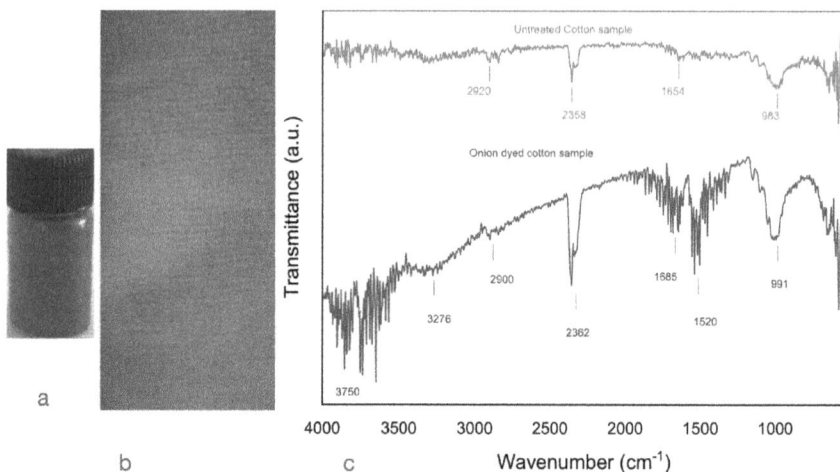

FIGURE 5.8 Onion peel nanoemulsion: (a) nanoemulsion, (b) dyed cotton, (c) FTIR graph of dyed and undyed cotton.

polyester-cotton blends was observed post-dyeing. The reduction in tear strength was 8.0–11.2% in cotton and 6.7–8.3% in polyester-cotton. The onion peel nano-emulsion–dyed cotton (72–75%) exhibited higher antibacterial activity than the polyester-cotton blended fabric (55–65%). The dyed fabric exhibited antibacterial activity post-25 home washes in cotton (18–31%) and polyester-cotton (19–23%) fabric against *S. aureus* and *E. coli* [50].

5.6 CONCLUSION

Research and technology for fruit and vegetable biomass collection, handling, and utilization for sustainable textile practices are available, but the applications are still in a nascent stage. Utilization of agricultural processing waste to impart technical finishes is advantageous for the health of producers, end users, and the environment. The abundant availability of agricultural by-products ensures their economic potential for use in technical finishes on textiles. Production of moth repellents, dye, and finishes sourced from plant parts and mud can be started by common people at the village level in small- to medium-scale units. The natural finishes presented in the chapter provided multiple advantages on fabric surfaces, including ultraviolet protection, flame retardancy, water resistance, and antimicrobial characteristics. Encouraging results of bacterial reduction (*E. coli* and *S. aureus*) were achieved with peanut and chickpea testa on textiles, verifying their use in various technical textiles, particularly in medical textiles. The technology mentioned in the chapter is economical and will certainly help to sustain the small- and medium-scale textile technical finishes industries.

REFERENCES

1. Konda A, Prakash A, Moss GA, Schmoldt M, Grant GD, Guha S. 2020. Aerosol filtration efficiency of common fabrics used in respiratory cloth masks. ACS Nano 14(5):6339–6347. https://doi.org/10.1021/acsnano.0c03252
2. Beesoon S, Behary N, Perwuelz A. 2020. Universal masking during COVID-19 pandemic: Can textile engineering help public health? Narrative review of the evidence. Preventive Medicine 139:106236. https://doi.org/10.1016/j.ypmed.2020.106236
3. Rahman MM, Koh J, Hong KH. 2022. Coloration and multi-functionalization of cotton fabrics using different combinations of aqueous natural plant extracts of onion peel, turmeric root, and pomegranate rind. Industrial Crops and Products 188:115562. https://doi.org/10.1016/j.indcrop.2022.115562
4. Xu W, Xu L, Pan H, Wang L, Shen Y. 2022. Superamphiphobic cotton fabric with photocatalysis and ultraviolet shielding property based on hierarchical ZnO/halloysite nanotubes hybrid particles. Colloids and Surfaces A: Physicochemical and Engineering Aspects 654:129995. https://doi.org/10.1016/j.colsurfa.2022.129995
5. Lee DT, Jamir JD, Peterson GW, Parsons GN. 2020. Protective fabrics: Metal-organic framework textiles for rapid photocatalytic sulfur mustard simulant detoxification. Matter 2(2):404–415. https://doi.org/10.1016/j.matt.2019.11.005
6. Pandey R, Pandit P, Pandey S, Mishra S. 2020. Solutions for sustainable fashion and textile industry. In Pintu Pandit et al. (eds), Recycling from Waste in Fashion and Textiles: A Sustainable & Circular Economic Approach. Scrivener Publishing, Beverly, MA, 33–72. https://doi.org/10.1002/9781119620532.ch3
7. Zambrano MC, Pawlak JJ, Daystar J, Ankeny M, Venditti RA. 2021. Impact of dyes and finishes on the aquatic biodegradability of cotton textile fibers and microfibers released on laundering clothes: Correlations between enzyme adsorption and activity and biodegradation rates. Marine Pollution Bulletin 165:112030. https://doi.org/10.1016/j.marpolbul.2021.112030
8. Buscio V, Crespi M, Gutiérrez-Bouzán C. 2014. A critical comparison of methods for the analysis of indigo in dyeing liquors and effluents. Materials 7(9):6184–6193. https://doi.org/10.3390/ma7096184
9. Wahyuningsih S, Ramelan AH, Wardani DK, Aini FN, Sari PL, Tamtama BPN, Kristiawan YR. 2017. Indigo dye derived from Indigofera tinctoria as natural food colorant. IOP Conference Series: Materials Science and Engineering 193(1):012048.
10. Pandey R, Patel S, Pandit P, Shanmugam N, Jose S. 2018. Colouration of textiles using roasted peanut skin—an agro processing residue. Journal of Cleaner Production 172:1319–1326. https://doi.org/10.1016/j.jclepro.2017.10.268
11. Seiko J, Pandit P, Pandey R. 2019. Chickpeahusk—A potential agro waste for coloration and functional finishing of textiles. Industrial Crops & Products 142:111833. https://doi.org/10.1016/j.indcrop.2019.111833
12. Pandit P, Jose S, Pandey R. 2020. Groundnut Testa: A potential agro residue for single bath dyeing and protective finishing of cotton fabric. Waste and Biomass Valorization 12(6):3383–3394. https://doi.org/10.1007/s12649-020-01214-y
13. Lugoloobi I, Memon H, Akampumuza O, Balilonda A. 2020. Advanced physical applications of modified cotton. In H Wang, H Memon (eds) Cotton Science and Processing Technology, Textile Science and Clothing Technology. Springer Nature, Singapore. https://doi.org/10.1007/978-981-15-9169-3_18
14. El-Zawahry MM, El Khatib HS, Shokry GM, Rashad HG. 2022. One-pot robust dyeing of cotton fabrics with multifunctional chamomile flower dyes. Fibers and Polymers 23(8):2234–2249. https://doi.org/10.1007/s12221-022-4206-0

15. Rahmati N, Eslahi N, Rashidi A. 2022. Antibacterial finishing of cotton with silk hydrogel and chamomile extract. Journal of Cleaner Production 336:130465.
16. Zhang, Y, Zhou Q, Xia W, Rather LJ, Li Q. 2022. Sonochemical mordanting as a green and effective approach in enhancing cotton bio natural dye affinity through soy surface modification. Journal of Cleaner Production 336:130465. https://doi.org/10.1016/j.jclepro.2022.130465
17. Rattanaphani S, Chairat M, Bremner JB, Rattanaphani V. 2007. An adsorption and thermodynamic study of lac dyeing on cotton pretreated with chitosan. Dyes Pigments 72(1):88–96. https://doi.org/10.1021/acssuschemeng.9b06928
18. Pisitsak P, Hutakamol J, Thongcharoen R, Phokaew P, Kanjanawan K, Saksaeng N. 2016. Improving the dyeability of cotton with tannin-rich natural dye through pretreatment with whey protein isolate. Industrial Crops & Products 79:47–56. https://doi.org/10.1016/j.indcrop.2015.10.043
19. Pisitsak P, Tungsombatvisit N, Singhanu K. 2018. Utilization of waste protein from Antarctic krill oil production and natural dye to impart durable UV-properties to cotton textiles. Journal of Cleaner Production 174:1215–1223. https://doi.org/10.1016/j.jclepro.2017.11.010
20. Haji A. 2020. Plasma activation and chitosan attachment on cotton and wool for improvement of dyeability and fastness properties. Pigment Resin Technology 49(6):483–489. https://doi.org/10.1108/PRT-02-2020-0017
21. Ticha MB, Haddar W, Meksi N, Guesmi A, Mhenni MF. 2016. Improving dyeability of modified cotton fabrics by the natural aqueous extract from red cabbage using ultrasonic energy. Carbohydrate Polymers 154:287–295. https://doi.org/10.1016/j.carbpol.2016.06.056
22. Alaysuy O, Snari RM, Alfi AA, Aldawsari AM, Abu-Melha S, Khalifa ME, El-Metwaly NM. 2022. Development of green and sustainable smart biochromic and therapeutic bandage using red cabbage (Brassica oleracea L. Var. capitata) extract encapsulated into alginate nanoparticles. International Journal of Biological Macromolecules 211:390–399. https://doi.org/10.1016/j.ijbiomac.2022.05.062
23. Sandra N, Alessandro P. 2021. Consumers' preferences, attitudes and willingness to pay for bio-textile in wood fibers. Journal of Retailing and Consumer Services 58:102304. https://doi.org/10.1016/j.jretconser.2020.102304
24. Kalia S, Thakur K, Celli A, Kiechel MA, Schauer CL. 2013. Surface modification of plant fibers using environment friendly methods for their application in polymer composites, textile industry and antimicrobial activities: A review. Journal of Environmental Chemical Engineering 1(3):97–112 https://doi.org/10.1016/j.jece.2013.04.009
25. Shabbir M, Rather LJ, Mohammad F. 2018. Economically viable UV-protective and antioxidant finishing of wool fabric dyed with Tagetes erecta flower extract: Valorization of marigold. Industrial Crops and Products 119:277–282 https://doi.org/10.1016/j.indcrop.2018.04.016
26. Claro A, Melo MJ, de Melo JSS, Van den Berg KJ, Burnstock A, Montague M, Newman R. 2010. Identification of red colorants in van Gogh paintings and ancient Andean textiles by microspectrofluorimetry. Journal of Cultural Heritage 11(1):27–34. https://doi.org/10.1016/j.culher.2009.03.006
27. Szostek B, Orska-Gawrys J, Surowiec I, Trojanowicz M. 2003. Investigation of natural dyes occurring in historical Coptic textiles by high-performance liquid chromatography with UV–Vis and mass spectrometric detection. Journal of Chromatography A 1012(2):179–192. https://doi.org/10.1016/S0021-9673(03)01170-1
28. Jose S, Thomas S, Pandit P, Pandey R. 2023. Handbook of Museum Textiles, Volume 2: Scientific and Technological Research. John Wiley & Sons, Beverly, MA.

29. Samanta AK. 2020. Bio-dyes, bio-mordants and bio-finishes: Scientific analysis for their application on textiles. In Chemistry and Technology of Natural and Synthetic Dyes and Pigments. IntechOpen. https://doi.org/10.5772/intechopen.92601

30. Prabha R, Vasugi RN. 2012. Filarial repellent finish using medicinal plants. Indian Journal Science 1(1):74–76.

31. Basak S, Samanta KK, Saxena S, Chattopadhyay SK, Narkar R, Mahangade R, Hadge GB. 2015. Flame resistant cellulosic substrate using banana pseudostem sap. Polish Journal of Chemical Technology 17(1):123–133. http://dx.doi.org/10.1515%2Fpjct-2015-0018

32. Pan Y, Yang X, Xu M, Sun G. 2017. Preparation of mud-coated silk fabrics with antioxidant and antibacterial properties. Materials Letters 191:10–13. https://doi.org/10.1016/j.matlet.2016.12.123

33. Sivakumar V, Vijaeeswarri J, Anna JL. 2011. Effective natural dye extraction from different plant materials using ultrasound. Industrial Crops and Products 33(1):116–122. https://doi.org/10.1016/j.indcrop.2010.09.007

34. Gangwar AKS, Singh MK, Vishnoi P, Shakyawar DB, Maity S. 2022. Sodium lignosulfonate: An industrial bio-waste for the colouration and UV protective finish of nylon fabric. Fibres & Textiles in Eastern Europe 1(151):77–85.

35. Sarkar AK. 2004. An evaluation of UV protection imparted by cotton fabrics dyed with natural colorants. BMC Dermatology 4(15):1–15.

36. Benelli G, Rajeswary M, Vijayan P, Senthilmurugan S, Alharbi NS, Kadaikunnan S, Khaled JM, Govindarajan M. 2018. Boswellia ovalifoliolata (Burseraceae) essential oil as an eco-friendly larvicide? Toxicity against six mosquito vectors of public health importance, non-target mosquito fishes, backswimmers, and water bugs. Environmental Science and Pollution Research 25(11):10264–10271. https://doi.org/10.1007/s11356-017-8820-0

37. Islam MT, Liman MLR, Roy MN, Hossain MM, Repon MR, Mamun MAA. 2021. Cotton dyeing performance enhancing mechanism of mangiferin enriched bio-waste by transition metals chelation. The Journal of The Textile Institute 1–13. https://doi.org/10.1080/00405000.2021.1892337

38. Venkatraman PD, Sayed U, Parte S, Korgaonkar S. 2022. Novel antimicrobial finishing of organic cotton fabrics using nano-emulsions derived from Karanja and Gokhru plants. Textile Research Journal. https://doi.org/10.1177/00405175221113364

39. Vajpayee M, Dave H, Singh M, Ledwani L. 2022. Cellulase enzyme based wet-pretreatment of lotus fabric to improve antimicrobial finishing with A. indica extract and enhance natural dyeing: Sustainable approach for textile finishing. ChemistrySelect 7(25):e202200382. https://doi.org/10.1002/slct.202200382

40. Ben Sassi A, Ascrizzi R, Chiboub W, Cheikh Mhamed A, ElAyeb A, Skhiri F, Tounsi Saidani M, Mastouri M, Flamini G. 2022. Volatiles, phenolic compounds, antioxidant and antibacterial properties of kohlrabi leaves. Natural Product Research 36(12):3143–3148. https://doi.org/10.1080/14786419.2021.1940177

41. Yadav I, Juneja SK, Chauhan S. 2015. Temple waste utilization and management: A review. International Journal of Engineering Technology Science and Research 2:14–19.

42. Rehan M, Ibrahim GE, Mashaly HM, Hasanin M, Rashad HG, Mowafi S. 2022. Simultaneous dyeing and multifunctional finishing of natural fabrics with Hibiscus flowers extract. Journal of Cleaner Production 133992. https://doi.org/10.1016/j.jclepro.2022.133992

43. Jia Y, Jiang H, Wang Y, Liu Z, Liang P. 2022. Fabrication of bio-based coloristic and ultraviolet protective cellulosic fabric using chitosan derivative and chestnut shell extract. Fibers and Polymers 23(10):2760–2768.

44. Sadeghi-Kiakhani M, Tehrani-Bagha AR, Gharanjig K, Hashemi E. 2019. Use of pomegranate peels and walnut green husks as the green antimicrobial agents to reduce the consumption of inorganic nanoparticles on wool yarns. Journal of Cleaner Production 231:1463–1473.

45. Liman MLR, Islam MT, Hossain M, Sarker P. 2020. Sustainable dyeing mechanism of polyester with natural dye extracted from watermelon and their UV protective characteristics. Fibers and Polymers 21(10):2301–2313. https://doi.org/10.1007/s12221-020-1135-7

46. Sun SS, Tang RC. 2011. Adsorption and UV protection properties of the extent form honeysuckle into wool. Industrial & Engineering Chemistry Research Journal 50(8):4217–4224.

47. Blackburn RS. 2017. Natural dyes in madder (Rubia spp.) and their extraction and analysis in historical textiles. Coloration Technology 133(6):449–462. https://doi.org/10.1111/cote.12308

48. Jiang Q, Li P, Liu Y, Zhu P. 2022. Flame retardant cotton fabrics with anti-UV properties based on tea polyphenol-melamine-phenylphosphonic acid. Journal of Colloid and Interface Science. https://doi.org/10.1016/j.jcis.2022.09.084

49. Teli MD, Pandit P. 2018. Development of thermally stable and hygienic colored cotton fabric made by treatment with natural coconut shell extract. Journal of Industrial Textiles 48(1):87–118 https://doi.org/10.1177%2F1528083717725113

50. Joshi S, Kambo N, Dubey S, Shukla P, Pandey R. 2021. Effect of onion (Allium cepa L.) peel extract-based nanoemulsion on anti-microbial and UPF properties of cotton and cotton blended fabrics. Journal of Natural Fibers. https://doi.org/10.1080/15440478.2021.1964127

51. Sreenivasan VK, Zvyagin AV, Goldys EM. 2013. Luminescent nanoparticles and their applications in the life sciences. Journal of Physics: Condensed Matter 25(19):194101.

52. Singh P, Borthakur A, Singh R, Awasthi S, Pal DB, Srivastava P, Tiwary D, Mishra PK. 2017. Utilization of temple floral waste for extraction of valuable products: A close loop approach towards environmental sustainability and waste management. Pollution 3(1):39–45.

53. Kumar, V, Kumari S, Kumar P. 2020. Management and sustainable energy production using flower waste generated from temples. Environmental Degradation: Causes and Remediation Strategies 1:154.

54. Niranjan A, Prakash D. 2008. Chemical constituents and biological activities of turmeric (Curcuma longa l.)-a review. Journal of Food Science and Technology 45(2):109–116.

55. Yang C, Wang Z, Ou C, Chen M, Wang L, Yang Z. 2014. A supramolecular hydro gelator of curcumin. Chemical Communications 50(66):9413–9415. https://doi.org/10.1039/C4CC03139C

56. Habib N, Ali A, Adeel S, Aftab M, Inayat A. 2022. Assessment of wild turmeric–based eco-friendly yellow natural bio-colorant for dyeing of wool fabric. Environmental Science and Pollution Research 1–12. https://doi.org/10.1007/s11356-022-22450-w

57. Lim SH, Hudson SM. 2004. Application of a fiber-reactive chitosan derivative to cotton fabric as an antimicrobial textile finish. Carbohydrate Polymers 56(2):227–234. https://doi.org/10.1016/j.carbpol.2004.02.005

58. Ahmed OK, El-Bendary MA, Elsayed NA. 2022. Creation of a functional cotton headcover (turban/bonnet) via lac and turmeric natural dyes. International Design Journal 12(6):221–232. https://dx.doi.org/10.21608/idj.2022.267382

59. Rahman MM, Kim M, Youm K, Kumar S, Koh J, Hong KH. 2023. Sustainable one-bath natural dyeing of cotton fabric using turmeric root extract and chitosan biomordant. Journal of Cleaner Production 382:135303. https://doi.org/10.1016/j.jclepro.2022.135303

60. Otaviano BTH, Sannomiya M, Da Costa SA, Da Costa SM, De Lima FS, Tangerina MMP, Tamayose CI, Ferreira MJP, Freeman HS, Flor JP. 2022. Pomegranate peel extract and zinc oxide as a source of natural dye and functional material for textile fibers aiming for photoprotective properties. Materials Chemistry and Physics 126766. https://doi.org/10.1016/j.matchemphys.2022.126766

61. Da Silva MG, De Barros MAS, De Almeida RTR, Pilau EJ, Pinto E, Soares G, Santos JG. 2018. Cleaner production of antimicrobial and anti-UV cotton materials through dyeing with eucalyptus leaves extract. Journal of Cleaner Production 199:807–816. https://doi.org/10.1016/j.jclepro.2018.07.221

62. Nguyen HL, Bechtold T. 2021. Thermal stability of natural dye lakes from Canadian Goldenrod and onion peel as sustainable pigments. Journal of Cleaner Production 315:128195. https://doi.org/10.1016/j.jclepro.2021.128195

63. Dhiman J, Kaith BS. 2020. Fabrication of high performance biodegradable Holarrhena antidysenterica fiber based adsorption devices. Arabian Journal of Chemistry 13(12):8734–8749. https://doi.org/10.1016/j.arabjc.2020.10.004

64. Pandey R, Prasad GK, Dubey A, Arputhraj A, Raja ASM, Sinha MK, Jose S. 2022. Tellicherry bark microfiber: Characterization and processing. Journal of Natural Fibers 19(16):13288–13299. http://dx.doi.org/10.1080/15440478.2022.2089432

65. Cherdcho W, Nithettham S, Charoenpanich J. 2019. Removal of Cr(VI) from synthetic wastewater by adsorption onto coffee ground and mixed waste tea. Chemosphere 221:758–767.

66. Tian D, Xu Z, Zhang D, Chen W, Cai J, Deng H, Sun Z, Zhou Y. 2019. Micro–mesoporous carbon from cotton waste activated by $FeCl_3/ZnCl_2$: Preparation, optimization, characterization and adsorption of methylene blue and eriochrome black T. Journal of Solid State Chemistry 269:580–587.

67. Yusop MFM, Aziz A, Ahmad MA. 2022. Conversion of teak wood waste into microwave-irradiated activated carbon for cationic methylene blue dye removal: Optimization and batch studies. Arabian Journal of Chemistry 15(9):104081. https://doi.org/10.1016/j.arabjc.2022.104081

68. Qiao H, Zhou Y, Yu F, Wang E, Min Y, Huang Q, Pang L, Ma T. 2015. Effective removal of cationic dyes using carboxylate-functionalized cellulose nanocrystals. Chemosphere 141:297–303. https://doi.org/10.1016/j.chemosphere.2015.07.078

69. Kamsonlian S, Shukla B. 2013. Optimization of process parameters using response surface methodology (RSM): Removal of Cr (VI) from aqueous solution by wood apple shell activated carbon (WASAC). Research Journal of Chemical Sciences 606.

70. Bhaumik M, Choi HJ, Seopela MP, McCrindle RI, Maity A. 2014. Highly effective removal of toxic Cr (VI) from wastewater using sulfuric acid-modified avocado seed. Industrial & Engineering Chemistry Research 53(3):1214–1224.

71. Peláez-Cid AA, Romero-Hernández V, Herrera-González AM, Bautista-Hernández A, Coreño-Alonso O. 2020. Synthesis of activated carbons from black sapote seeds, characterization and application in the elimination of heavy metals and textile dyes. Chinese Journal of Chemical Engineering 28(2):613–623. https://doi.org/10.1016/j.cjche.2019.04.021

72. Srivastava VC, Mall ID, Mishra IM. 2006. Equilibrium modelling of single and binary adsorption of cadmium and nickel onto bagasse fly ash. Chemical Engineering Journal 117(1):79–91.

73. Giraldo-Gutiérrez L, Moreno-Piraján JC. 2008. Pb (II) and Cr (VI) adsorption from aqueous solution on activated carbons obtained from sugar cane husk and sawdust. Journal of Analytical and Applied Pyrolysis 81(2):278–284.

74. Sadaf S, Bhatti HN. 2014. Batch and fixed bed column studies for the removal of Indosol Yellow BG dye by peanut husk. Journal of the Taiwan Institute of Chemical Engineers 45:541e553 https://doi.org/10.1016/j.jtice.2013.05.004

75. Liu W, Hao Y, Jiang J, Zhu A, Zhu J, Dong Z. 2016. Production of a bioflocculant from pseudomonas veronii L918 using the hydrolyzate of peanut hull and its application in the treatment of ash-flushing wastewater generated from coal fired power plant. Bioresource Technology 218:318e325 https://doi.org/10.1016/j.biortech.2016.06.108

76. Zvinowanda CM, Okonkwo JO, Sekhula MM, Agyei NM, Sadiku R. 2009. Application of maize tassel for the removal of Pb, Se, Sr, U and V from borehole water contaminated with mine wastewater in the presence of alkaline metals. Journal of Hazardous Materials 164(2–3):884–891. https://doi.org/10.1016/j.jhazmat.2008.08.110

77. Edokpayi JN, Odiyo JO, Msagati TA, Popoola EO. 2015. A novel approach for the removal of lead (II) ion from wastewater using mucilaginous leaves of diceriocaryum eriocarpum plant. Sustainability 7(10):14026–14041. https://doi.org/10.3390/su71014026

78. Aniagor CO, Menkiti MC. 2018. Kinetics and mechanistic description of adsorptive uptake of crystal violet dye by lignified elephant grass complexed isolate. Journal of Environmental Chemical Engineering 6(2):2105–2118. https://doi.org/10.1016/j.jece.2018.01.070

79. Vinayagam R, Quadras M, Varadavenkatesan T, Debraj D, Goveas LC, Samanth A, Balakrishnan D, Selvaraj R. 2023. Magnetic activated carbon synthesized using rubber fig tree leaves for adsorptive removal of tetracycline from aqueous solutions. Environmental Research 216:114775. https://doi.org/10.1016/j.envres.2022.114775

80. Abudi ZN, Lattieff FA, Chyad TF, Abbas MN. 2018. Isotherm and kinetic study of the adsorption of different dyes from aqueous solution using banana peel by two different methods. Global Journal of Bio-science and Biotechnology 7(2):220–229.

81. Amari A, Alalwan B, Eldirderi MM, Mnif W, Rebah FB. 2019. Cactus material-based adsorbents for the removal of heavy metals and dyes: A review. Materials Research Express 7(1):012002. DOI 10.1088/2053-1591/ab5f32

82. Shakya A, Núñez-Delgado A, Agarwal T. 2019. Biochar synthesis from sweet lime peel for hexavalent chromium remediation from aqueous solution. Journal of Environmental Management 251:109570. https://doi.org/10.1016/j.jenvman.2019.109570

83. Yunus ZM, Al-Gheethi A, Othman N, Hamdan R, Ruslan NN. 2020. Removal of heavy metals from mining effluents in tile and electroplating industries using honey-dew peel activated carbon: A microstructure and techno-economic analysis. Journal of Cleaner Production 251:119738. https://doi.org/10.1016/j.jclepro.2019.119738

84. Bakircioglu Y, Bakircioglu D, Akman S. 2003. Solid phase extraction of bismuth and chromium by rice husk. Journal of Trace and Microprobe Techniques 21(3):467–478. https://doi.org/10.1081/TMA-120023063

85. Ahmaruzzaman M, Gupta VK. 2011. Rice husk and its ash as low-cost adsorbents in water and wastewater treatment. Industrial & Engineering Chemistry Research 50(24):13589–13613. https://doi.org/10.1021/ie201477c

86. Moradi P, Hayati S, Ghahrizadeh T. 2020. Modeling and optimization of lead and cobalt biosorption from water with Rafsanjan pistachio shell, using experiment based models of ANN and GP, and the grey wolf optimizer. Chemometrics and Intelligent Laboratory Systems 202:104041. https://doi.org/10.1016/j.chemolab.2020.104041

87. Aziz, A, Khan MNN, Yusop MFM, Jaya EMJ, Jaya MAT, Ahmad MA. 2021. Single-stage microwave-assisted coconut-shell-based activated carbon for removal of dichlorodiphenyltrichloroethane (DDT) from aqueous solution: Optimization and batch studies. International Journal of Chemical Engineering. https://doi.org/10.1155/2021/9331386

88. Mohadi R, Palapa NR, Taher T, Siregar PMSBN, Juleanti N, Wijaya A, Lesbani A. 2021. Removal of Cr (VI) from aqueous solution by biochar derived from rice husk. Communications in Science and Technology 6(1):11–17. https://doi.org/10.21924/cst.6.1.2021.293

89. Liu Y, Shan H, Pang Y, Zhan H, Zeng C. 2022. Iron modified chitosan/coconut shell activated carbon composite beads for Cr (VI) removal from aqueous solution. International Journal of Biological Macromolecules 224(1):156–169. https://doi.org/10.1016/j.ijbiomac.2022.10.112

90. Yusop MFM, Jaya EMJ, Ahmad MA. 2022. Single-stage microwave assisted coconut shell based activated carbon for removal of Zn (II) ions from aqueous solution–optimization and batch studies. Arabian Journal of Chemistry 15(8):104011. https://doi.org/10.1016/j.arabjc.2022.104011

91. Rao ML, Savithramma N. 2011. Biological synthesis of silver nanoparticles using Svensonia Hyderabadensis leaf extract and evaluation of their antimicrobial efficacy. Journal of Pharmaceutical Sciences and Research 3(3):1117.

92. Rajakumar G, Gomathi T, Thiruvengadam M, Rajeswari VD, Kalpana VN, Chung IM. 2017. Evaluation of anti-cholinesterase, antibacterial and cytotoxic activities of green synthesized silver nanoparticles using from Millettia pinnata flower extract. Microbial Pathogenesis 103:123–128. https://doi.org/10.1016/j.micpath.2016.12.019

93. Ibrahim NA, Eid BM, Abdellatif FH. 2018. Advanced materials and technologies for antimicrobial finishing of cellulosic textiles. In Handbook of Renewable Materials for Coloration and Finishing. Scrivener Publishing, Beverly, MA, 301–356.

94. Prabhu N, Raj DT, Yamuna GK, Ayisha SS, Puspha J, Innocent D. 2010. Synthesis of silver phyto nanoparticles and their antibacterial efficacy. Digest Journal of Nanomaterials & Biostructures 5(1):185–189.

95. Rosace G, Migani V, Colleoni C, Massafra MR, Sancaktaroglu E. 2010. Nanoparticle finishes influence on color matching of cotton fabrics. In AK Haghi (ed.) Chemistry and Chemical Engineering Research Progress. Nova Science Publishers Inc., New York, 29–44.

96. Roy P, Das B, Mohanty A, Mohapatra S. 2017. Green synthesis of silver nanoparticles using Azadirachta indica leaf extract and its antimicrobial study. Applied Nanoscience 7(8):843–850. https://doi.org/10.1007/s13204-017-0621-8

97. Abdellatif FHH, Abdellatif MM, Ahmed HM. 2022. Advanced nanotechnologies for finishing of cellulosic textiles. In Sharma et al. (eds) Fundamentals of Nano–Textile Science. Apple Academic Press, New York, 73–99. https://doi.org/10.1201/9781003277316

98. Kumar BS. 2016. Study on antimicrobial effectiveness of sliver nano coating over cotton fabric through green approach. International Journal of Pharma Science & Research 7:363–368.

99. Ahmed S, Ahmad M, Swami BL, Ikram S. 2016. A review on plants extract mediated synthesis of silver nanoparticles for antimicrobial applications: A green expertise. Journal of Advanced Research 7(1):17–28. https://doi.org/10.1016/j.jare.2015.02.007

100. Dubas ST, Kumlangdudsana P, Potiyaraj P. 2006. Layer-by-layer deposition of antimicrobial silver nanoparticles on textile fibers. Colloids and Surfaces A: Physicochemical and Engineering Aspects 289(1–3):105–109. https://doi.org/10.1016/j.colsurfa.2006.04.012

101. Ramteke C, Chakrabarti T, Sarangi BK, Pandey RA. 2013. Synthesis of silver nanoparticles from the aqueous extract of leaves of Ocimum sanctum for enhanced antibacterial activity. Journal of Chemistry. https://doi.org/10.1155/2013/278925
102. Kumara Swamy M, Sudipta KM, Jayanta K, Balasubramanya S. 2015. The green synthesis, characterization, and evaluation of the biological activities of silver nanoparticles synthesized from Leptadenia reticulata leaf extract. Applied Nanoscience 5(1):73–81. https://doi.org/10.1007/s13204-014-0293-6
103. Yugandhar P, Haribabu R, Savithramma N. 2015. Synthesis, characterization and antimicrobial properties of green-synthesised silver nanoparticles from stem bark extract of Syzygium alternifolium (Wt.) Walp. 3 Biotech 5:1031–1039. https://doi.org/10.1007/s13205-015-0307-4
104. Shateri-Khalilabad M, Yazdanshenas ME, Etemadifar A. 2017. Fabricating multifunctional silver nanoparticles-coated cotton fabric. Arabian Journal of Chemistry 10:S2355–S2362. https://doi.org/10.1016/j.arabjc.2013.08.013
105. Mohanta YK, Panda SK, Syed A, Ameen F, Bastia AK, Mohanta TK. 2018. Bio-inspired synthesis of silver nanoparticles from leaf extracts of Cleistanthus collinus (Roxb.): Its potential antibacterial and anticancer activities. IET Nanobiotechnology 12(3):343–348. https://doi.org/10.1049/iet-nbt.2017.0203
106. Thiruchelvi R, Jayashree P, Mirunaalini K. 2021. Synthesis of silver nanoparticle using marine red seaweed Gelidiella acerosa-a complete study on its biological activity and its characterisation. Materials Today: Proceedings 37:1693–1698. https://doi.org/10.1016/j.matpr.2020.07.242
107. Wang Y, Wang X, Li T, Ma P, Zhang S, Du M, Dong W, Xie Y, Chen M. 2018. Effects of melanin on optical behavior of polymer: From natural pigment to materials applications. ACS Applied Materials & Interfaces 10(15):13100–13106. https://doi.org/10.1021/acsami.8b02658
108. Krishnaraj C, Jagan EG, Rajasekar S, Selvakumar P, Kalaichelvan PT, Mohan NJCSBB. 2010. Synthesis of silver nanoparticles using Acalypha indica leaf extracts and its antibacterial activity against water borne pathogens. Colloids and Surfaces B: Biointerfaces 76(1):50–56. https://doi.org/10.1016/j.colsurfb.2009.10.008
109. Khan SA, Lee CS. 2020. Green biological synthesis of nanoparticles and their biomedical applications. In Inamuddin, Asiri (eds) Applications of Nanotechnology for Green Synthesis. Springer, Cham, 247–280. https://doi.org/10.1007/978-3-030-44176-0_10
110. Ilić V, Šaponjić Z, Vodnik V, Potkonjak B, Jovančić P, Nedeljković J, Radetić M. 2009. The influence of silver content on antimicrobial activity and color of cotton fabrics functionalized with Ag nanoparticles. Carbohydrate Polymers 78(3):564–569 https://doi.org/10.1016/j.carbpol.2009.05.015
111. Jain D, Daima HK, Kachhwaha S, Kothari SL. 2009. Synthesis of plant-mediated silver nanoparticles using papaya fruit extract and evaluation of their anti microbial activities. Digest Journal of Nanomaterials and Biostructures 4(3):557–563.
112. Gokarneshan N, Velumani K. 2018. Significant trends in nano finishes for improvement of functional properties of fabrics. In Mohd Yusuf (ed.) Handbook of Renewable Materials for Coloration and Finishing. Scrivener Publishing, Beverly, MA, 387–434.
113. Savithramma N, Rao ML, Rukmini K, Devi PS. 2011. Antimicrobial activity of silver nanoparticles synthesized by using medicinal plants. International Journal of ChemTech Research 3(3):1394–1402.
114. Shameli K, Ahmad MB, Zargar M, Yunus WMZW, Ibrahim NA, Shabanzadeh P, Moghaddam MG. 2011. Synthesis and characterization of silver/montmorillonite/chitosan bionanocomposites by chemical reduction method and their antibacterial activity. International Journal of Nanomedicine 6:271–284. https://doi.org/10.2147%2FIJN.S16043

115. Vivek M, Kumar PS, Steffi S, Sudha S. 2011. Biogenic silver nanoparticles by Gelidiella acerosa extract and their antifungal effects. Avicenna Journal of Medical Biotechnology 3(3):143.

116. Wang X, Gao W, Xu S, Xu W. 2012. Luminescent fibers: In situ synthesis of silver nanoclusters on silk via ultraviolet light-induced reduction and their antibacterial activity. Chemical Engineering Journal 210:585–589. https://doi.org/10.1016/j.cej.2012.09.034

117. Sasikala A, Linga Rao M, Savithramma N, Prasad TNVKV. 2015. Synthesis of silver nanoparticles from stem bark of Cochlospermum religiosum (L.) Alston: An important medicinal plant and evaluation of their antimicrobial efficacy. Applied Nanoscience 5(7):827–835. https://doi.org/10.1007/s13204-014-0380-8

118. Beg M, Maji A, Mandal AK, Das S, Aktara MN, Jha PK, Hossain M. 2017. Green synthesis of silver nanoparticles using Pongamia pinnata seed: Characterization, antibacterial property, and spectroscopic investigation of interaction with human serum albumin. Journal of Molecular Recognition 30(1):e2565. https://doi.org/10.1002/jmr.2565

119. Lee JH, Lim JM, Velmurugan P, Park YJ, Park YJ, Bang KS, Oh BT. 2016. Photobiologic-mediated fabrication of silver nanoparticles with antibacterial activity. Journal of Photochemistry and Photobiology B: Biology 162:93–99. https://doi.org/10.1016/j.jphotobiol.2016.06.029

120. Radetić M. 2013. Functionalization of textile materials with silver nanoparticles. Journal of Materials Science 48(1):95–107. https://doi.org/10.1007/s10853-012-6677-7.

121. Radetić M, Marković D. 2019. Nano-finishing of cellulose textile materials with copper and copper oxide nanoparticles. Cellulose 26:8971–8991. https://doi.org/10.1007/s10570-019-02714-4

122. Shaheen TI, El-Naggar ME, Abdelgawad AM, Hebeish A. 2016. Durable antibacterial and UV protections of in situ synthesized zinc oxide nanoparticles onto cotton fabrics. International Journal of Biological Macromolecules 83:426–432. https://doi.org/10.1016/j.ijbiomac.2015.11.003

123. Murugan K, Senthilkumar B, Senbagam D, Al-Sohaibani S. 2014. Biosynthesis of silver nanoparticles using Acacia leucophloea extract and their antibacterial activity. International Journal of Nanomedicine 9:2431. https://doi.org/10.2147%2FIJN.S61779

124. Logaranjan K, Raiza AJ, Gopinath SC, Chen Y, Pandian K. 2016. Shape-and size-controlled synthesis of silver nanoparticles using Aloe vera plant extract and their antimicrobial activity. Nanoscale Research Letters 11(1):520. https://doi.org/10.1186/s11671-016-1725-x

125. Veerasamy R, Xin TZ, Gunasagaran S, Xiang TFW, Yang EFC, Jeyakumar N, Dhanaraj SA. 2011. Biosynthesis of silver nanoparticles using mangosteen leaf extract and evaluation of their antimicrobial activities. Journal of Saudi Chemical Society 15(2):113–120. https://doi.org/10.1016/j.jscs.2010.06.004

126. Krithiga N, Rajalakshmi A, Jayachitra A. 2015. Green synthesis of silver nanoparticles using leaf extracts of Clitoria ternatea and Solanum nigrum and study of its antibacterial effect against common nosocomial pathogens. Journal of Nanoscience 1–8. http://dx.doi.org/10.1155/2015/928204

127. Akintelu SA, Folorunso AS, Oyebamiji AK, Erazua EA. 2019. Antibacterial potency of silver nanoparticles synthesized using Boerhaavia diffusa leaf extract as reductive and stabilizing agent. International Journal of Pharma Sciences & Research 10(12):374–380.

128. Chandraker SK, Lal M, Dhruve P, Yadav AK, Singh RP, Varma RS, Shukla R. 2022. Bioengineered and biocompatible silver nanoparticles from Thalictrum foliolosum DC and their biomedical applications. Clean Technologies and Environmental Policy 24:2479–2494. https://doi.org/10.1007/s10098-022-02329-7

129. Siddiqi KS, Husen A, Rao RA. 2018. A review on biosynthesis of silver nanoparticles and their biocidal properties. Journal of Nanobiotechnology 16(1):1–28. https://doi.org/10.1186/s12951-018-0334-5

130. Qiao SZ, Liu J, Lu GQM. 2011. Synthetic chemistry of nanomaterials. In R Xu et al. (eds) Modern Inorganic Synthetic Chemistry. Elsevier, 479–506. https://doi.org/10.1016/B978-0-444-53599-3.10021-6

131. Nierstrasz VA. 2009. Enzyme surface modification of textiles. In Surface Modification of Textiles. Woodhead Publishing, 139–163. https://doi.org/10.1533/9781845696689.139

132. Akin DE, Foulk JA, Dodd RB, McAlister III DD. 2001. Enzyme-retting of flax and characterization of processed fibers. Journal of Biotechnology 89(2–3):193–203. https://doi.org/10.1016/S0168-1656(01)00298-X

133. Chen J, Wang Q, Hua Z, Du G. 2007. Research and application of biotechnology in textile industries in China. Enzyme and Microbial Technology 40(7):1651–1655. https://doi.org/10.1016/j.enzmictec.2006.07.040

134. Pandey R. 2016. Fiber extraction from dual-purpose flax. Journal of Natural Fibers 13(5):565–577. https://doi.org/10.1080/15440478.2015.1083926

135. Mahmoodi NM, Arami M, Mazaheri F, Rahimi S. 2010. Degradation of sericin (degumming) of Persian silk by ultrasound and enzymes as a cleaner and environmentally friendly process. Journal of Cleaner Production 18(2):146–151. https://doi.org/10.1016/j.jclepro.2009.10.003

136. Ferrero F, Periolatto M. 2012. Glycerol in comparison with ethanol in alcohol-assisted dyeing. Journal of Cleaner Production 33:127–131. https://doi.org/10.1016/j.jclepro.2012.04.018

137. Monier M, El-Sokkary AMA. 2012. Modification and characterization of cellulosic cotton fibers for efficient immobilization of urease. International Journal of Biological Macromolecules 51(1–2):18–24. https://doi.org/10.1016/j.ijbiomac.2012.04.019

138. Parvinzadeh M. 2007. Effect of proteolytic enzyme on dyeing of wool with madder. Enzyme and Microbial Technology 40(7):1719–1722. https://doi.org/10.1016/j.enzmictec.2006.10.026

139. Balan DS, Monteiro RT. 2001. Decolorization of textile indigo dye by ligninolytic fungi. Journal of Biotechnology 89(2–3):141–145. https://doi.org/10.1016/S0168-1656(01)00304-2

140. Pandey R, Mishra S, Dubey R. 2023. Luxurious Sustainable Fibers. In SS Muthu (ed.) Novel Sustainable Raw Material Alternatives for the Textiles and Fashion Industry. Springer Nature Switzerland, Cham, 57–79. https://doi.org/10.1007/978-3-031-37323-7_4

141. Katkar MM. 2022. Cowdung: A bio coating. In Proceedings: Fourth International Conference on Advances in Materials Science (ICAMS-2020), January 20–21, 2020.

142. Mukherjee G, Ghosh S. 2020. Use of cow dung as mosquito repellant. International Research Journal of Pharmacy and Medical Sciences 3(1):61–62.

143. Yang C, Wang Z, Ou C, Chen M, Wang L, Yang Z. 2014. A supramolecular hydrogelator of curcumin. Chemical Communications 50(66):9413–9415. https://doi.org/10.1039/C4CC03139C

144. Pandey R. 2021. Extraction of flax fibers by using gel retting method (Patent No. 372965).

145. Malucelli G. 2018. Bio-macromolecules: A new flame retardant finishing strategy for textiles. In Mohd Yusuf (ed.) Handbook of Renewable Materials for Coloration and Finishing. Scrivener Publishing, Beverly, MA, 357–386.

146. Feng Y, Zhou Y, Li D, He S, Zhang F, Zhang G. 2017. A plant-based reactive ammonium phytate for use as a flame-retardant for cotton fabric. Carbohydrate Polymers 175:636–644. https://doi.org/10.1016/j.carbpol.2017.06.129

147. Saleemi S, Naveed T, Riaz T, Memon H, Awan JA, Siyal MI, Xu F, Bae J. 2020. Surface functionalization of cotton and PC fabrics using SiO_2 and ZnO nanoparticles for durable flame retardant properties. Coatings 10(2):124. https://doi.org/10.3390/coatings10020124

148. Lam YL, Kan CW, Yuen CWM. 2011. Flame-retardant finishing In cotton fabrics using zinc oxide co-catalyst. Journal of Applied Polymer Science 121:612–621. https://doi.org/10.1002/app.33738

149. Guido E, Colleoni C, De Clerck K, Plutino MR, Rosace G. 2014. Influence of catalyst in the synthesis of a cellulose-based sensor: Kinetic study of 3-glycidoxypropyltrimethoxysilane epoxy ring opening by Lewis acid. Sensors and Actuators B: Chemical 203:213–222. https://doi.org/10.1016/j.snb.2014.06.126

150. Abou-Okeil A, El-Sawy SM, Abdel-Mohdy FA. 2013. Flame retardant cotton fabrics treated with organophosphorus polymer. Carbohydrate Polymers 92(2):2293–2298. https://doi.org/10.1016/j.carbpol.2012.12.008

151. Wei Z, Yang Z, Chen Z, Yu T, Li Y. 2022. A green and facile strategy to enhance thermal stability and flame retardancy of unidirectional flax fabric based on fully biobased system. Industrial Crops and Products 188:115610. https://doi.org/10.1016/j.indcrop.2022.115610

152. Zain G, Jordanov I, Bischof S, Magovac E, Šišková AO, Vykydalová A, Kleinová A, Mičušík M, Mosnáčková K, Nováčiková J, Mosnáček J. 2022. Flame-retardant finishing of cotton fabric by surface-initiated photochemically induced atom transfer radical polymerization. Cellulose 1–22. https://doi.org/10.1007/s10570-022-04982-z

153. Saleem MA, Nazir A, Nazir F, Ayaz P, Faizan MQ, Usman M, Hussain T. 2019. Comparison of UV protection properties of cotton fabrics treated with aqueous and methanolic extracts of Solanum nigrum and Amaranthus viridis plants. Photodermatology, Photoimmunology & Photomedicine 35(2):93–99. https://doi.org/10.1111/phpp.12427

154. Rungruangkitkrai N, Mongkholrattanasit R, Phoophat P, Chartvivatpornchai N, Sirimungkararat S, Wongkasem K, Tuntariyanond P, Nithithongsakol N, Chollakup R. 2020. UV-protection property of Eri silk fabric dyed with natural dyes for eco-friendly textiles. IOP Conference Series: Materials Science and Engineering 773(1):012027.

155. Agnhage T, Zhou Y, Guan J, Chen G, Perwuelz A, Behary N, Nierstrasz V. 2017. Bioactive and multifunctional textile using plant-based madder dye: Characterization of UV protection ability and antibacterial activity. Fibers and Polymers 18(11):2170–2175. https://doi.org/10.1007/s12221-017-7115-x

156. Kumar SS. 2007. Biopolymers in medical applications. Technical Textiles, CRC Press, Boca Raton, FL, 1–15.

157. Teli MD, Sheikh J, Pradhan C. 2014. Simultaneous natural dyeing and antibacterial finishing using chitosan from bio-waste. Melliand International 20(3):171–172.

158. Wang M, Yi N, Fang K, Zhao Z, Xie R, Chen W. 2022. Deep colorful antibacterial wool fabrics by high-efficiency pad dyeing with insoluble curcumin. Chemical Engineering Journal 139121. https://doi.org/10.1016/j.cej.2022.139121

159. Yuranova T, Rincon AG, Bozzi A, Parra S, Pulgarin C, Albers P, Kiwi J. 2003. Antibacterial textiles prepared by RF-plasma and vacuum-UV mediated deposition of silver. Journal of Photochemistry and Photobiology A: Chemistry 161(1):27–34.

160. El-Rafie MH, Mohamed AA, Shaheen TI, Hebeish A. 2010. Antimicrobial effect of silver nanoparticles produced by fungal process on cotton fabrics. Carbohydrate Polymers 80(3):779–782. https://doi.org/10.1016/j.carbpol.2009.12.028

161. Cushnie TPT, Lamb AJ. 2005. Antimicrobial activity of flavonoids. International Journal of Antimicrobial Agents 26(5):343–356. https://doi.org/10.1016/j.ijantimicag.2005.09.002

162. Mishra S, Pandey R, Singh MK. 2016. Development of sanitary napkin by flax carding waste as absorbent core with herbal and antimicrobial efficiency. International Journal of Science Environment and Technology 5(2):404–411.

163. Kumar H, Bhardwaj K, Sharma R, Nepovimova E, Kuča K, Dhanjal DS, Verma R, Bhardwaj P, Sharma S, Kumar D. 2020. Fruit and vegetable peels: Utilization of high value horticultural waste in novel industrial applications. Molecules 25(12):2812. https://doi.org/10.3390/molecules25122812

164. Li J, He J, Huang Y. 2017. Role of alginate in antibacterial finishing of textiles. International Journal of Biological Macromolecules 94:466–473. https://doi.org/10.1016/j.ijbiomac.2016.10.054

165. Qiu Q, Chen S, Li Y, Yang Y, Zhang H, Quan Z, Qin X, Wang R, Yu J. 2020. Functional nanofibers embedded into textiles for durable antibacterial properties. Chemical Engineering Journal 384:123241. https://doi.org/10.1016/j.cej.2019.123241

166. Jose S, Nachimuthu S, Das S, Kumar A. 2018. Moth proofing of wool fabric using nano kaolinite. The Journal of the Textile Institute 109(2):225–231. https://doi.org/10.1080/00405000.2017.1336857

167. Ren G, Song Y, Li X, Wang B, Zhou Y, Wang Y, Ge B, Zhu X. 2018. A simple way to an ultra-robust superhydrophobic fabric with mechanical stability, UV durability, and UV shielding property. Journal of Colloid and Interface Science 522:57–62. https://doi.org/10.1016/j.jcis.2018.03.038

168. Shao Z, Wang Y, Bai H. 2020. A superhydrophobic textile inspired by polar bear hair for both in air and underwater thermal insulation. Chemical Engineering Journal 397:125441. https://doi.org/10.1016/j.cej.2020.125441

169. Xue CH, Chen J, Yin W, Jia ST, Ma JZ. 2012. Superhydrophobic conductive textiles with antibacterial property by coating fibers with silver nanoparticles. Applied Surface Science 258(7):2468–2472. https://doi.org/10.1016/j.apsusc.2011.10.074

170. Cheng Y, Zhu T, Li S, Huang J, Mao J, Yang H, Gao S, Chen Z, Lai Y. 2019. A novel strategy for fabricating robust superhydrophobic fabrics by environmentally-friendly enzyme etching. Chemical Engineering Journal 355:290–298.

171. Tissera ND, Wijesena RN, Perera JR, de Silva KN, Amaratunge GA. 2015. Hydrophobic cotton textile surfaces using an amphiphilic graphene oxide (GO) coating. Applied Surface Science 324:455–463.

172. Gould P. 2003. Textiles gain intelligence. Materials Today 6(10):38–43. https://doi.org/10.1016/S1369-7021(03)01028-9

173. Kaynak A, Najar SS, Foitzik RC. 2008. Conducting nylon, cotton and wool yarns by continuous vapor polymerization of pyrrole. Synthetic Metals 158(1–2):1–5.

174. Li P, Du D, Guo L, Guo Y, Ouyang J. 2016. Stretchable and conductive polymer films for high-performance electromagnetic interference shielding. Journal of Materials Chemistry C 4(27):6525–6532.

175. Song J, Yang W, Liu X, Zhang W, Zhang Y. 2016. ASA/graphite/carbon black composites with improved EMI SE, conductivity and heat resistance properties. Iranian Polymer Journal 25(2):111–118.

176. Chung DDL. 2001. Electromagnetic interference shielding effectiveness of carbon materials. Carbon 39(2):279–285.

177. Shen B, Zhai W, Zheng W. 2014. Ultrathin flexible graphene film: An excellent thermal conducting material with efficient EMI shielding. Advanced Functional Materials 24(28):4542–4548. https://doi.org/10.1002/adfm.201400079

178. He Q, Lv J, Xu H, Zhang L, Zhong Y, Sui X, Wang B, Chen Z, Mao Z. 2019. Enhancing electrical conductivity and electrical stability of polypyrrole-coated cotton fabrics via surface microdissolution. Journal of Applied Polymer Science 136(21):47515. https://doi.org/10.1002/app.47515

179. Lide DR. 1995. CRC Handbook of Chemistry and Physics. 76th ed. CRC Press, Boca Raton, FL. https://doi.org/10.1101/2021.02.28.433265

180. Sun C, Zhao J, Guo Z, Zhao Y, Cai Z, Ge F. 2019. A novel method to fabricate nitrogen and oxygen co-doped flexible cotton-based electrode for wearable supercapacitors. ChemElectroChem 6(15):4049–4058. https://doi.org/10.1002/celc.201901123

181. Islam MZ, Dong Y, Khoso NA, Ahmed A, Deb H, Zhu Y, Wentong Y, Fu Y. 2019. Continuous dyeing of graphene on cotton fabric: Binder-free approach for electromagnetic shielding. Applied Surface Science 496:143636. https://doi.org/10.1016/j.apsusc.2019.143636

182. Yazdanshenas ME, Shateri-Khalilabad M. 2013. In situ synthesis of silver nanoparticles on alkali-treated cotton fabrics. Journal of Industrial Textiles 42(4):459–474. https://doi.org/10.1177/1528083712444297

183. Hassabo AG, El-Naggar ME, Mohamed AL, Hebeish AA. 2019. Development of multifunctional modified cotton fabric with tri-component nanoparticles of silver, copper and zinc oxide. Carbohydrate Polymers 210:144–156. https://doi.org/10.1016/j.carbpol.2019.01.066

184. Atwa Y, Maheshwari N, Goldthorpe IA. 2015. Silver nanowire coated threads for electrically conductive textiles. Journal of Materials Chemistry C 3(16):3908–3912. https://doi.org/10.1039/C5TC00380F

185. Kim JS, Kuk E, Yu KN, Kim JH, Park SJ, Lee HJ, Kim SH, Park YK, Park YH, Hwang CY Kim YK. 2007. Antimicrobial effects of silver nanoparticles. Nanomedicine: Nanotechnology, Biology and Medicine 3(1):95–101. https://doi.org/10.1016/j.nano.2006.12.001

186. Verma NK, Conroy J, Lyons PE, Coleman J, O'Sullivan MP, Kornfeld H, Kelleher D, Volkov Y. 2012. Autophagy induction by silver nanowires: A new aspect in the biocompatibility assessment of nanocomposite thin films. Toxicology and Applied Pharmacology 264(3):451–461. https://doi.org/10.1016/j.taap.2012.08.023

187. Song MJ, Hwang SW, Whang D. 2010. Amperometric hydrogen peroxide biosensor based on a modified gold electrode with silver nanowires. Journal of Applied Electrochemistry 40(12):2099–2105. https://doi.org/10.1007/s10800-010-0191-x

188. Abdel-Halim ES, Abdel-Mohdy FA, Al-Deyab SS, El-Newehy MH. 2010. Chitosan and monochlorotriazinyl-β-cyclodextrin finishes improve antistatic properties of cotton/polyester blend and polyester fabrics. Carbohydrate Polymers 82(1):202–208. https://doi.org/10.1016/j.carbpol.2010.04.077

189. Ali SW, Bairagi S, Shankar P. 2020. Biomaterials-based nanogenerator: Futuristic solution for integration into smart textiles. In Frontiers of Textile Materials: Polymers, Nanomaterials, Enzymes, and Advanced Modification Techniques, 189–201. https://doi.org/10.1002/9781119620396.ch9

190. Agarwal P, Yusuf M, Khan SA, Prasad L. 2018. Bio-colorants as photosensitizers for dye sensitized solar cell (DSSC). In Handbook of Renewable Materials for Coloration and Finishing. Scrivener Publishing, Beverly, MA, 279–300.

191. Boyo AO, Abdulsalami IO, Oluwa T, Oluwole SO, Umar A. 2013. Development of dye sensitized solar cells using Botuje green leaves (Jathopha Curcas Linn). Science Journal of Physics 2013:1–4.

192. Barbosa R, Gupta SK, Srivastava, BB, Villarreal A, De Leon H, Peredo M, Bose S, Lozano K. 2021. Bright and persistent green and red light-emitting fine fibers: A potential candidate for smart textiles. Journal of Luminescence 231:117760. https://doi.org/10.1016/j.jlumin.2020.117760

193. Leem JW, Choi SH, Kim, SR, Kim SW, Choi KH, Kim YL. 2017. Scalable and continuous nanomaterial integration with transgenic fibers for enhanced photoluminescence. Materials Horizons 4(2):281–289. https://doi.org/10.1039/C6MH00423G

194. El-Nagar K, Saudy MA, Eatah AI, Masoud MM. 2006. DC pseudo plasma discharge treatment of polyester textile surface for disperse dyeing. Journal of the Textile Institute 97(2):111–117. https://doi.org/10.1533/joti.2005.0169

195. El-Zawahry MM, Ibrahim NA, Eid MA. 2006. The impact of nitrogen plasma treatment upon the physical-chemical and dyeing properties of wool fabric. Polymer-Plastics Technology and Engineering 45(10):1123–1132. https://doi.org/10.1080/03602550600728943

196. Kalia S, Kaith BS, Kaur I. 2009. Pretreatments of natural fibers and their application as reinforcing material in polymer composites—A review. Polymer Engineering & Science 49(7):1253–1272. https://doi.org/10.1002/pen.21328

197. Nithya E, Radhai R, Rajendran R, Shalini S, Rajendran V, Jayakumar S. 2011. Synergetic effect of DC air plasma and cellulase enzyme treatment on the hydrophilicity of cotton fabric. Carbohydrate Polymers 83(4):1652–1658. https://doi.org/10.1016/j.carbpol.2010.10.027

198. Lee HW, Lu Y, Zhang Y, Fu C, Huang D. 2021. Physicochemical and functional properties of red lentil protein isolates from three origins at different pH. Food Chemistry 358:129749. https://doi.org/10.1016/j.foodchem.2021.129749

199. Jain P, Nimbrana S, Kalia G. 2010. Antimicrobial activity and phytochemical analysis of Eucalyptus tereticornis bark and leaf methanolic extracts. International Journal of Pharmaceutical Sciences Review and Research 4:126–128.

200. Godstime O, Enwa O, Augustina J, Christopher E. 2014. Mechanism of antimicrobial actions of phytochemicals against enteric pathogens-a review. Journal of Pharmaceutical, Chemical and Biological Sciences 2:77–85.

201. Ammayappan L, Jose S. 2015. Functional aspects, ecotesting, and environmental impact of natural dyes. In SS Muthu (ed.), Handbook of Sustainable Apparel Production. CRC Press, Boca Raton, FL, 333–350.

202. Soulef K, Yahia A. 2017. Antibacterial effects of glycosides extract of Glycyrrhizaglabra L. from the region of Djamaa (South East of Algeria). Journal of Medicinal Herbs and Ethnomedicine 3:22–25.

6 Nanofibers for Technical Textile Applications

6.1 INTRODUCTION

Nanotechnology products and procedures are first used in the textile manufacturing business. Hence, nanotechnology branding in the fabric sector will be crucial for its widespread. The technical textile industry stands to benefit significantly from nanotechnology, presenting a significant commercial opportunity. To fully utilize the enhanced properties of greater surface area and reduced weight, research in fiber spinning should prioritize the development of nanoscale fibers. A long-term research proposal is being considered to create a new technology for coating textile fabrics with nanofibers [1, 2]. This technology aims to utilize nanolayer self-assembly as an advanced surface modification technique to develop a foundational framework for the process. Mastering the fundamentals of this process will enable the commercialization of nanofiber-coated textiles into low-cost, high-value technical textile products. Nanotechnology-coated textiles enable the creation of intelligent structures on a nano or submicron scale, resulting in stronger and fundamentally different fibers that offer improved water, oil, and soil repellency; UV/IR protection; improved electric conductivity; protective barrier properties against biological and chemical agents; antimicrobial performance; protective shielding; self-healing; thermal and flame protection; and high mechanical and abrasion resistance. US military researchers have developed a nanotechnology-based chemical that can coat fabric, resulting in antibacterial and water- and stain-resistant properties, making it suitable for hospital bedding, healthcare uniforms, and air conditioning filters. A nozzle-less spinning approach has been devised to get around the problems with standard electrospinning setups' lack of productivity and difficulty producing webs of consistent quality. This method creates an electrically charged jet of polymer fluid by using a high-voltage power source, a spinneret electrode, and a collecting electrode. This approach produces a highly uniform deposition of nanofibers on fabric without altering the functional qualities of the substrate and has the ability to use a higher-voltage electric field (up to 80 KVA compared to the typical 10 KV). Furthermore, it enables the fusion of high-performance materials, producing a coated composite structure. These coated nanofibrous webs have potential for use as a membrane layer in various technical textile applications, making this novel electrospinning coating technology suitable for high-tech engineering applications [1, 3].

6.2 ATTRACTIVE FEATURES OF NANOFIBERS

Nanofibers possess unique functional properties that are attributed to their extremely high surface-to-weight ratio compared to standard fibers. These properties include effective barrier properties against microorganisms and fine particulates; potential for strong mechanical properties per unit weight; low energy requirements for production; a high pore volume with small pores; a high surface area to mass ratio; and specific finishing effects like smoothing, high-cover, and stain repellency. Additionally, nanofibers have high surface energy, which results in good moisture vapor transmission rates. Coating and functionalizing webs onto textile substrates can create functional fabrics that are lighter in weight and offer a wider range of features compared to traditional coated fabrics [4].

6.3 PRINCIPLES

The foundation of electrospinning is the more than 400-year-old electrostatic attraction theory. William Gilbert found in 1600 that an electrostatic attraction between opposing charges could cause a water droplet to be pulled near rubbed amber and take the shape of a cone. According to Coulomb's law, which states that opposite charges attract and like charges repel, this attraction occurs (Figure 6.1).

Therefore, it is the electrostatic attraction between opposite charges that causes the deformation of solution droplets into a nanorange (50–900 nm) fibrous structural network. Since the work of Zeleny [a.3] and Taylor [a.4], substantial process model advancements have been accomplished. Many of the previous models have

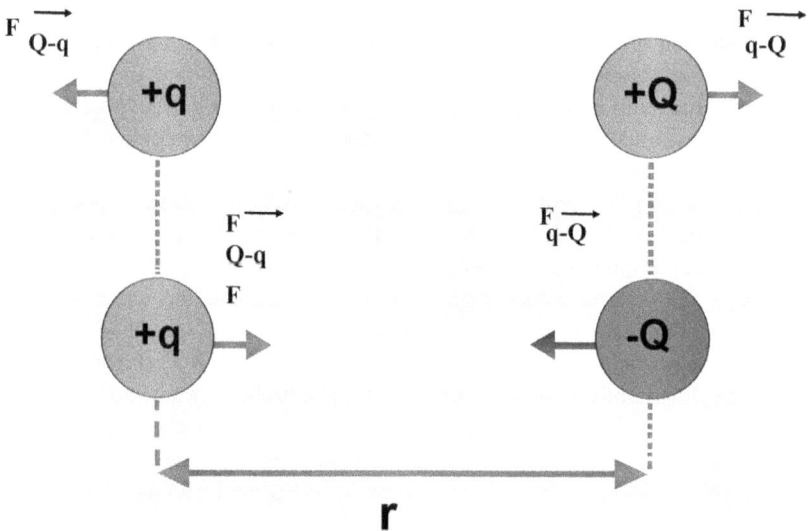

FIGURE 6.1   A schematic diagram of the electrospinning principle.

been tested against data from experiments and utilized to help explain certain experimental findings [3].

It is observed that a nanofibrous network consists of three stages:

1. Jet initiation,
2. Jet thinning, and
3. Fluid instabilities.

6.3.1 Jet Initiation

Jet initiation occurs when a fluid drop comes into contact with a conductor (spinning electrode) and is exposed to a strong electric field (Figure 6.2).

This results in the formation of a Taylor cone structure. Research on jet initiation has focused on the process of cone formation, which has been observed to have a semi-vertical angle of approximately 49.3° for all conducting fluids, as described by Taylor (Figure 6.3).

6.3.2 Jet Thinning

After departing from the Taylor cone, the jet experiences forces and must adhere to the principle of conservation of mass, causing it to gradually decrease in size until it reaches the collector electrode. This reduction in size occurs in two distinct stages. The first stage is characterized by the jet thinning as a straight line, while the second stage involves thinning due to bending instability.

Nanofibrous Coated Web Formation via droplets

FIGURE 6.2 A schematic diagram of electrospinning web formation via nozzle spinning process and technology.

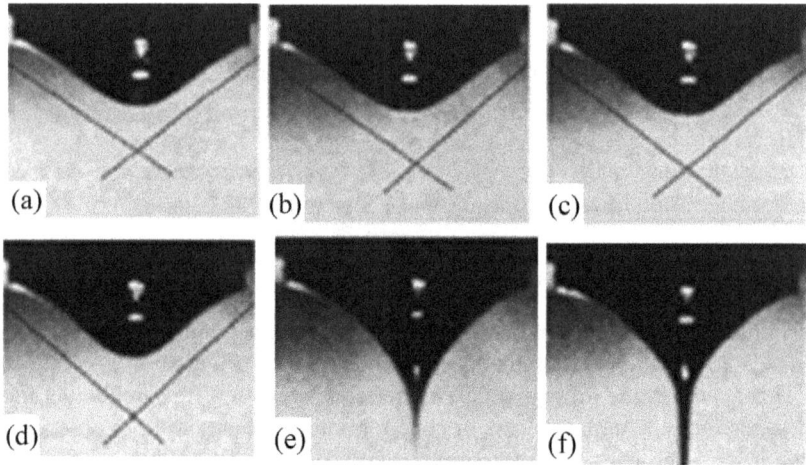

FIGURE 6.3   Micrograph of electrospinning Taylor cone generation.

FIGURE 6.4   Bending instability (whipping action) of electrospinning process.

6.3.3 FLUID INSTABILITIES

Electrospraying is a process where droplet formation occurs due to either to high or too-low force attraction, resulting in the absence of a fibrous structure. However, with an optimal force of attraction, fluid instabilities occur through a bending or "whipping" instability action that determines the quality and thickness of the fibers [5] (Figure 6.4).

Y.M. Shin and colleagues demonstrated that increasing the potential difference and thus the strength of the electric field during electrospinning can result in improved crystallinity if given enough flight time (time of whipping action). This might be a result of the jet being stretched further by the stronger electric field. As an alternative, the electric field might encourage polymer molecule alignment, similar to electrophoresis, resulting in more aligned molecules under a stronger field.

6.4   ELECTROSPINNING PROCESS

Figure 6.1 depicts a continuous stream of nanofibers made using the nozzle-less coating technique, a type of electrospinning without nozzles. Figure 6.5 illustrates a schematic diagram that outlines the process and technology utilized in the electrospinning of nozzles [3].

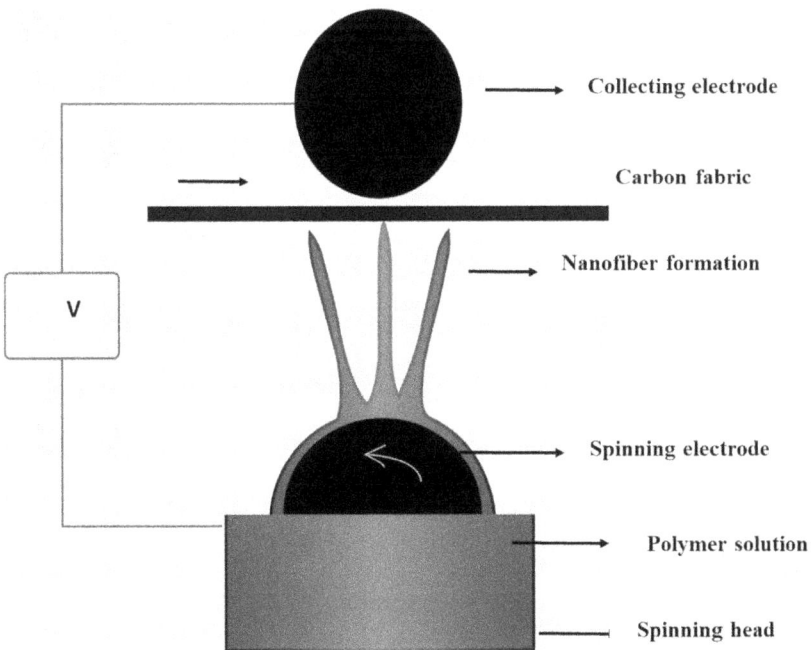

FIGURE 6.5   A schematic diagram of process and technology of nozzle electrospinning.

In this method, a coating of nanofibers is applied using a spinning head. The potential differential between a spinning electrode (+ve) and a collecting electrode (-ve) is what drives the process. As the potential difference between the electrodes grows, the polymer solution is extruded from the spinning electrode in the shape of a conical nanofibrous web. The spinning electrode is submerged in a polymer solution that is contained in the spinning head. This web is applied at a set rate (m/min) to a nonwoven fabric that is moving. The spinning electrode's rotation offers a steady supply of polymer solution for extrusion. The section that follows talks about the procedure's ideal parameters.

6.5 PROCESS PARAMETERS

Numerous factors have been found to influence the final characteristics of the electrospun fiber during electrospinning. As shown in Table 6.1, these variables can be divided into three categories: (1) characteristics of the solution used as the feedstock (solution parameters); (2) characteristics related to the design, geometry, and operation of the electrospinning apparatus; and (3) atmospheric conditions and other regional processing conditions (environmental parameters).

6.5.1 ROTATION OF SPINNING ELECTRODES

The number of revolutions per minute (RPM) at which the electrode rotates determines how quickly a layer of polymer solution forms on its surface. The ideal rotation speed for the spinning electrode depends on factors such as the type

TABLE 6.1

Electrospinning Process Parameters and Optimized Values on Nanospider Technology

Solution Parameters	Process Parameters	Environmental Conditions
Concentration ~10–20%	Electrostatic potential ~ 30 70 KV	Temperature ~18°C–30°C
Dynamic viscosity ~10–5000 mPas	Electric field strength < 0.62 kV/ mm	Relative humidity ~20–40%
Surface tension ~0.05–0.1 N/m	Electrostatic field shape	Atmospheric composition
Conductivity ~0.01–10 μS/cm	Distance between spinning and collection electrode ~12–17 cm	Local atmosphere flow
Solvent volatility boiling point ~80° –200 vapor pressure of a liquid at saturation vapor pressure of 0.35–10 kPa	Speed of collecting electrode (CE) m/min 0.13–1.56 m/min depending upon g/m^2 substate	Pressure
	Rotation of spinning electrode (SE) ~ 8–14 rpm	Air velocity
	Feed rate throughput 0.1–0.25 g/m^2 min[1]	

of electrode and polymer solution being used, as well as the throughput of the machine, which is measured in grams per square meter per minute.

6.5.2   DISTANCE OF ELECTRODES, VOLTAGE, AND ELECTRICAL FIELD INTENSITY

In order to compute the electrical field intensity (E), which is a critical component of electrospinning, one must divide the applied voltage (U) by the distance (D) between the spinning and collecting electrodes. The electrical field intensity is measured in kV/mm. While E can be used to compare product parameters such as the throughput of particular polymers, it is important to consider the distance between the electrodes in relation to the quality of the nanofiber layer produced. Using a shorter distance between the electrodes allows for lower voltage, but this can result in less space and time for the nanofibers to form, leading to less uniform layers with more defects. In the NS LAB Tool, the minimum distance between the spinning electrode and collector is 10 cm. Distances between 10 and 15 cm are recommended for water-soluble polymers (e.g., PVA) with high surface tension, while distances between 15 and 20 cm are more suitable for polymers dissolved in organic solvents. However, using very high distances above 20 cm should be avoided, as it can overload the HV power supplies and cause damage.

6.5.3   DISTANCE OF SPINNING AND COLLECTING ELECTRODE

Increasing the distance between the substrate material and the deposition nozzle (SE and CE) can often lead to a more uniform nanofiber layer due to increased width and improved throughput. But it's crucial to remember that the highest throughput is not always achieved at a distance of 40 mm, and the relationship between distance and homogeneity is not linear for all polymer solutions and substrates. Generally, for most polymer solutions and substrates, a distance of 120–150 mm tends to yield the best results.

6.5.4   POLYMERS

The molecular weight of a polymer can provide valuable information in determining its spinnability. Polymers with lower molecular weight tend to have lower dynamic viscosity in solution, allowing for higher concentrations of solid polymer in a solution. On the other hand, higher-molecular-weight polymers tend to have higher dynamic viscosity, requiring lower concentrations of polymer in a solution. However, there is no universal requirement for polymer molecular weight, as it can vary greatly, ranging from thousands to millions. It is important that the polymer can form a stable solution in at least one solvent.

6.5.4.1   Solvents and Spinning Solutions

The polymer solution properties are crucial in the electrospinning process, and each polymer requires an optimized spinning solution system. There is no general model or recipe that works well for all polymers, so polymer solutions are optimized empirically. It's important to prepare the polymer solution using at least one solvent without using harsh conditions such as rapidly increased heat and pressure. Solvent mixtures can be used to obtain optimal solution properties. Similar to the surface tension of the solvent, the surface tension of the produced polymer solution should be up to 0.05 N/m. The dynamic viscosity should be in the range of 60 to 7000 mPas, with most solutions falling between 100 and 3000 mPas. Viscosity tends to decrease as temperature increases since viscosity is a material inherent parameter. The conductivity of the solution should be between 0.01 μS/cm and 10 mS/cm, with most solutions falling between 0.01 μS/cm and 2 mS/cm. Conductivity can be increased by adding acids (for water solutions) or organic salts (such as quaternary ammonium salts).

6.5.5 Ambient Air Conditions

The environment inside the spinning chamber is a key factor in determining the caliber of the fibers generated as well as the operation's overall effectiveness. However, finding the optimal conditions for temperature and relative humidity can be a challenging task, as these parameters vary depending on the specific polymer system in use. As a result, the ideal conditions are typically determined through trial and error. To address this issue, it would be advantageous to incorporate an air conditioning unit into the NS LAB Tool, which could be used to regulate the temperature and relative humidity within the spinning chamber. This would allow for greater control over the process and potentially improve the quality of the fibers produced. Additionally, it has been observed that there is a strong correlation between relative humidity and machine throughput, as illustrated in the accompanying figures. Specifically, increasing the relative humidity tends to lead to a decrease in the basis weight of the fibers produced, regardless of the concentration of the spinning solution.

6.6 PRODUCT ANALYSIS AND CHARACTERIZATION

A sample of ultrafine nanocoated fabric is subjected to scientific analysis for quality control purposes, as well as functional evaluation, to determine its suitability for various end uses and specific requirements. The functional evaluation process involved testing the fabric's performance in different scenarios related to its intended use.

The following analytical techniques are used to characterize the properties of the samples.

6.6.1 Surface Morphology and Diameter

With the help of a Carl Zeiss EV 050 field emission scanning electron microscope (FESEM), images of the nanowebs were taken in several scanned locations in order to examine the surface morphology and diameter of the nanowebs. Using

Image J software (Fiji, Version IJ 1.46r), we examined 40 randomly selected fibers to calculate the average diameter of the nanofibers. On the basis of appearance, the density and evenness of the nanowebs are inferred. We also used an atomic force microscope (AFM; Agilent 5500 SPM AFM, Agilent Technologies, Inc., West Arizona, USA) to examine the nanofibrous web surface.

6.6.2 Optical Measurement Profiler

Scan (Nano Focus Inc.) uses a confocal sensor to evaluate the topography of the nanofibers and calculate the roughness index of the fibers formed in the webs.

6.6.3 Capillary Liquid Expulsion Porometry

By thoroughly wetting the porous material with a wetting soap liquid, the Model POROMETER 3G win 10.10 Series automatic capillary flow porometer is used to measure the material's pore diameters.

6.6.4 Tensile Properties

A universal tensile testing machine is used to measure the tensile properties of fibrous webs, including Young's modulus and tensile stress. The testing is done using a specified sample size 20 mm in length and 10 mm in width at a crosshead speed of 10 mm/min. Cushioned double-sided sticky paper tape is used to make sure the grips do not directly contact the fibrous web. Five samples are chosen at random and tested, and the mean values obtained are given.

6.6.5 Antibacterial Test

By being incubated in agar plates with gram-positive and gram-negative bacteria for 24 hours at 37°C using *Escherichia coli* (gram-negative) and *S. aureus* (gram-positive) bacteria, the electrospun nanofibers and fabric are put through an antibacterial test. After 24 hours, the standard test technique AATCC 100–2012 is used to calculate the percentage of reduction in bacterial colonies.

6.6.6 Bacterial Filtration

In a study on bacterial filtration, 15 cm between a six-stage cascade impactor and an aerosol chamber, a multilayered fabric and a nonwoven fabric with a nanoweb included in it are placed. A 10-micron nebulizer assembly is used to produce bacterial aerosol in the chamber, namely *Staphylococcus aureus*—ATCC6538. The aerosol is drawn through the test materials by a vacuum connected to the cascade impactor. The bacterial filtration efficiency (%BFE) was defined as the proportion of the bacterial aerosol count upstream compared to downstream [6].

6.6.7   Thermal Analysis

The thermal behavior and physical characteristics of the samples are measured using differential scanning calorimetry (DSC) and thermogravimetric analysis (TGA). A nitrogen environment was used for the TGA, which was carried out at a heating rate of 10°C/min at temperatures ranging from 30°C to 900°C. A temperature range of 30°C to 300°C and a heating rate of 10°C/min were used for the DSC experiment.

6.6.8   Viscosity Measurement

A Brookfield Viscometer II Pro is used to measure the viscosity of web solutions at a temperature of 25 °C.

6.6.9   Crystalline Structure

The crystalline structure of the materials is ascertained using wide-angle X-ray diffraction (XRD). Copper K-a radiation in reflectance mode is used to record the XRD traces over a range of 10° to 80° at a scanning speed of 2s/0.02°.

6.6.10   Chemical Analysis

The interaction of multi-walled carbon nanotubes (MWCNTs), polyacrylonitrile (PAN), and organic bio-polymeric materials is examined using Raman spectroscopy. The samples' Raman spectra are captured using a Raman and Micro PL system, with measurements made spanning the wavelength range of 4000–250 cm^{-1}.

6.6.11   Air Permeability

When there is a specified pressure difference between the fabric's two faces, the air permeability of a fabric is the rate at which air can move through a unit area of the fabric. The TEXTEST FX 3300 air permeability tester is used in accordance with the ASTM D737 standard to evaluate the air permeability of nanoweb-coated nonwoven fabric and multilayered fabric.

6.6.12   Ultraviolet Radiation Protection

The ultraviolet (UV) protection provided by nanowebs is evaluated using a UV-visible-NIR spectrophotometer. The fabrics that allowed the least amount of UV radiation transmission demonstrated a higher level of absorption and reflection, indicating superior UV radiation protection. The transmission percentage was then used to calculate an ultraviolet protection factor (UPF) rating.

6.6.13   Xenon Light Exposure

Researchers frequently expose materials to accelerated daytime settings using a xenon arc light tester to explore how materials deteriorate in the presence of light.

This device has a panel setting of 60.5% relative humidity and 21.3°C temperature, and it employs a xenon-arc burner. To conduct a test, five samples of a material are cut into 20 × 10 mm pieces and exposed to the xenon light source for 100 hours. The material's tensile properties are measured both before and after exposure to the light. This process allows researchers to study how exposure to light affects the material's strength and other mechanical properties.

6.6.14 WATER VAPOR TRANSMISSION

The upright cup method, or ASTM E 96 Procedure B, is used to assess the breathability or comfort of an adsorbent layer. This technique involves placing the adsorbent fabric in a cup with its coated side facing the water and exposing it to a conditioning chamber with a temperature of 20°C and a relative humidity of 50% for the duration of the night. The weight change (in grams), the time it took for the weight to change, and the test area (in square meters) expressed in grams per square meter per 24 hours are then used to calculate the water vapor transmission rate (WVTR).

6.7 APPLICATIONS OF ELECTROSPUN NANOFIBER-COATED WEBS

Nanofiber technology has opened up exciting possibilities for the development of innovative textile constructions that can be integrated into cutting-edge applications, especially in the realm of technical clothing, as depicted in Figure 6.6.

The production of nanofibers and their potential applications in a wide range of industries, including clean energy (such as solar cells, fuel cells, and batteries), electronics, healthcare (such as biomedical scaffolds and artificial organs), and the environment (such as filter membranes), have been the subject of numerous studies over the past ten years. This chapter focuses on potential applications that could be converted into high-value technical textiles with lighter weight and enhanced functionality compared to traditionally manufactured fabrics. When combined with nanotechnology coatings, these textiles have the potential to create smart and multifunctional fabrics that can be tailored for various technical textile applications. Research shows that nanofiber-based materials can offer a wide range of functional performance in defense applications, from body armor for ballistic and stab protection to thermal and chemical protection suits. For instance, composite armor can be used for ballistic and bomb protection, while fire protection suits can be created for firefighters. Furthermore, the use of photochromic and thermochromic dyes and the integration of antimicrobial and IR/UV absorption properties can further enhance the versatility of these textiles. In recent developments, researchers from North Carolina State University, (NCSU), USA, and National University of Singapore (NUS) have created a nanocoated fabric that can effectively kill or inactivate most viruses and bacteria, offering a reliable solution for chemical and biological warfare protection [7].

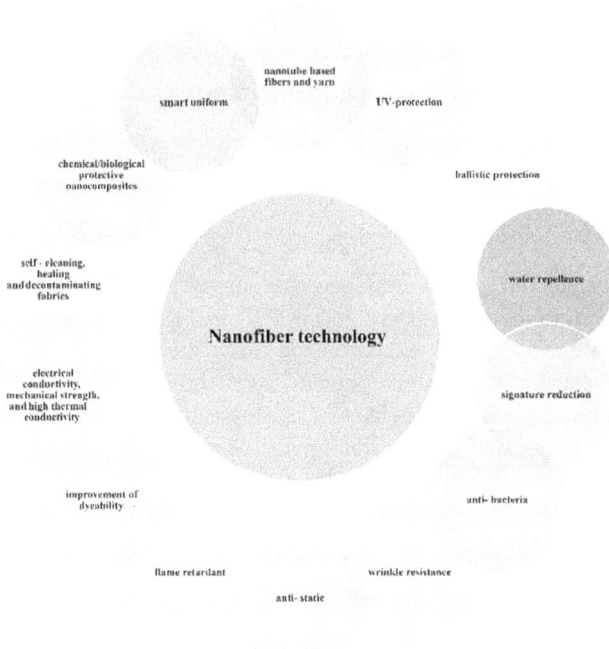

FIGURE 6.6 An application of nozzle electrospinning technology in technical textiles.

This fabric's nonwoven nanofiber structure makes it perfect for a variety of filtration and barrier applications against chemical and biological warfare agents because of its low density, high surface area to mass ratio, and tight pore size [8].

Additionally, this nanotechnology-based product offers superior moisture and vapor transportation while being up to 70% lighter than traditional charcoal-impregnated suits. Overall, the potential of nanofiber technology in defense applications is immense and can be used to develop a variety of forms and structures to accommodate the many requirements of situations involving modern warfare. The unique properties of these materials make them well suited for a diverse range of applications across multiple fields including filtration, anti-microbial, chemical resistance, bacterial filtration, UV-protective clothing, tissue engineering (specifically scaffolds), drug delivery, wound healing, and biomedical applications. Each of these applications will be briefly discussed in the following paragraphs.

6.7.1 HIGH-PERFORMANCE NANOWEBS FOR SUPERCAPACITOR APPLICATIONS

In recent times, there has been a significant demand for high-performance fibers that exhibit multifunctional characteristics. Electrospinning has emerged as a promising technique for generating ceramic nanofibers with various applications.

To create ceramic nanofibers, primarily oxide-based, two common processes, electrospinning and sol-gel, are typically combined. The sol-gel technique involves four primary steps, which include the formation of desired colloidal solutions from particles, followed by the deposition of the colloidal solution onto a substrate using methods such as spraying, coating, or electrospinning. The remaining organic or inorganic components are then pyrolyzed to create an amorphous or crystalline ceramic compound after the sol particles are allowed to gel in a state of continuous network. This method has been successful in synthesizing various oxide nanofibers and composites, including MgO, ZnO, CuO, NiO, TiO_2, SiO_2, Cr_2O_3, Al_2O_3, SnO_2, Fe_2O_3, $NiFe_2O_4$, $LiFePO_4$, and Pt. This method can also be used to create hybrid fibers, strontium ferrite composite nanofibers, and mixed Mg and Zn nanofibers. Recent research has shown that M-type strontium ferrite nanofibers can be successfully made using electrospinning and exhibit superior magnetic properties to those of sol-gel-produced nanoparticles, making them ideal for lighter-weight capacitor applications. However, the sol-gel technique has some drawbacks that haven't garnered much attention lately, such as poor bonding, a low degree of functional characteristics, limited wear resistance, and high permeability [9].

6.7.2   UV-PROTECTIVE NANOFIBROUS COATED WEBS

Due to their superior hydrophobicity and chemical stability brought on by the extremely electronegative properties of polyvinylidene difluoride (PVDF), nanofibers have thus far mostly been studied for usage as water filters and cell electrolyte membranes. However, less attention has been given to exploiting the inertness and chemical stability of PVDF for environmental protection purposes. In this section, we explore the use of PVDF's non-reactive properties and its functionalization with TiO_2 at various compositions to create webs with UV-protective properties for environmental degradation prevention. The newly developed coating layer has potential applications in the production of lightweight clothing for protection against UV radiation in various settings, including defense, paramilitary, and civilian contexts. It could also be used for creating tents, covers, shelter segments, combat suits, and camouflage nets for snowy environments [10].

6.7.3   BACTERIAL-RESISTANT AND BACTERIAL FILTRATION WEBS

Chitosan nanofiber-coated membranes have enormous potential as a layer for enhanced biomedical, antimicrobial, and filtration garments. Infiltration applications for textiles, such as protective gear for chemical, viral, and bacterial exposure as well as wound healing, can benefit from the web's inherent biological qualities. The production rate of the coated web is also significantly higher, at about 0.25 $g/m^2/min$, boosting the possibility of economic viability and enabling mass production [11].

6.7.4 NANOTUBE-BASED CONDUCTIVE WEBS

Nanocomposite fibers made of polyacrylonitrile with varying amounts of multi-walled carbon nanotubes were produced using a nozzleless high-throughput electrospinning method. These fibers were then coated onto nonwoven polypropylene fabric with success. The composite fibers exhibit percolation behavior that enhances their electrical conductivity, and they adhere to Ohm's law of conductivity. The surface conductivity of these fibers was determined to be on the order of 10^{-6} S/cm, indicating that they fall within the semi-conducting range [12].

6.7.5 MOISTURE MANAGEMENT WEBS

PAN was electrospun into nanofiber webs by the electrospinning process. The study examines the transport characteristics of a multi-layer textile that makes use of electrospun nanofiber mats as a permeable barrier fabric. Comparing a multi-layered fabric made from these nanofiber mats with a commercially available protective fabric called Gore-Tex (like PTFE), it was observed that using nanofiber membranes increased water vapor permeability while simultaneously providing outstanding windproof and adequate water-repellent qualities. In summary, the use of nanofiber membranes as opposed to coating materials offers several advantages, including enhanced wind proofing and water repellency, as well as improved water vapor permeability [1, 12].

6.7.6 FIRE RETARDANCY OF CLAY/CARBON NANOFIBER HYBRID WEB SHEETS IN FIBER COMPOSITES

The ability of nanowebs to resist fire was examined in a study. With 0.05 wt% and 0.20 wt% of Cloisite Na+ clay, the researchers produced hybrid sheets including carbon nanofibers and clay. A vacuum-assisted resin transfer molding procedure was used to integrate the sheets into the surface of laminated composites, which are comparable to conventional continuous fiber mats. The sheets were created utilizing a high-pressure filtration system. Cone calorimeter tests that exposed the laminated composites to a 50 kW/m^2 external radiant heat flux were used to assess their fire performance. The outcomes demonstrated that the laminated composites' fire behavior was superior and might potentially be helpful for air and automotive applications [2, 13].

6.7.7 PROTECTION AGAINST NUCLEAR, BIOLOGICAL, AND CHEMICAL AGENTS BY LAYERED NANOFIBROUS WEB

This research work focuses on electrospinning critical materials such as chitosan and Nylon 66 and functionalizing them with detoxifying agents to provide protection against chemical warfare agents. The study examines the morphology of the resulting nanoweb and its impact on defense and protective functionalities.

The morphology of the nanofibers has a significant impact on the functional and comfortable protective behaviors against nuclear, biological, and chemical (NBC) threats. The lightweight multilayered (multifunctional fabric) fabric developed in this study is highly beneficial, with various features that are advantageous for NBC protection [14, 15].

6.7.8   ULTRAVIOLET RADIATION-PROTECTIVE NANOFIBROUS WEB

The purpose of this study is to use fluorine's non-reactive characteristics and combine them with TiO_2 at various compositions to produce UV protective webs that are resistant to environmental degradation. The study involves the creation of defect-free PVDF nanofibrous webs and their subsequent coating onto polypropylene nonwoven textile substrates to create UV-protective fabrics. The research is focused on developing practical applications for high-tech UV-protective products, making it an application-oriented study [11].

6.7.9   BIOMEDICAL APPLICATIONS

At the Seoul National University laboratory, researchers are using electrospinning to produce scaffolds for tissue engineering from biodegradable and biocompatible synthetic or natural polymers. Additionally, magneto-rheological fluids are being electrospun to fabricate materials with energy-absorption properties [14, 16].

6.8   CASE STUDIES

6.8.1   CHITOSAN-COATED NANOFIBROUS WEB

The electrospinning coating of pure chitosan that had been dissolved in tetrachloro acetic acid (TFA) and dichloro methane (DCM) was the subject of a study [14]. With coating densities ranging from 0.25 to 1.2 gsm, polypropylene spun-bonded nonwoven fabric was applied to the created nanowebs. The ideal electrospinning solution conditions were identified and successfully established, producing a well-structured web with 11% electrospun chitosan in an electric field of 75 KV, 135 mm between spinning electrodes, and 5 rpm (throughput) spinning electrode rotation. A consistently coated nanofibrous web with a mean fiber diameter ranging from 1210 to 1221 nm was produced by repeating the electrocoating procedure under the same conditions but using different collector speeds. In contrast to AFM and FESEM photos, which exhibited fibrous disintegration at lower solution concentrations (2 and 5%) and poor web spinnability at higher concentrations (14%), AFM microphotographs of the produced chitosan web revealed an interconnected porous structure. Coating density had a significant impact on the spun web's shape [14] (Figure 6.7).

Due to their special characteristics, these nanofibrous coated membranes could be used in filtration, biomedical, and antimicrobial garments. Additionally,

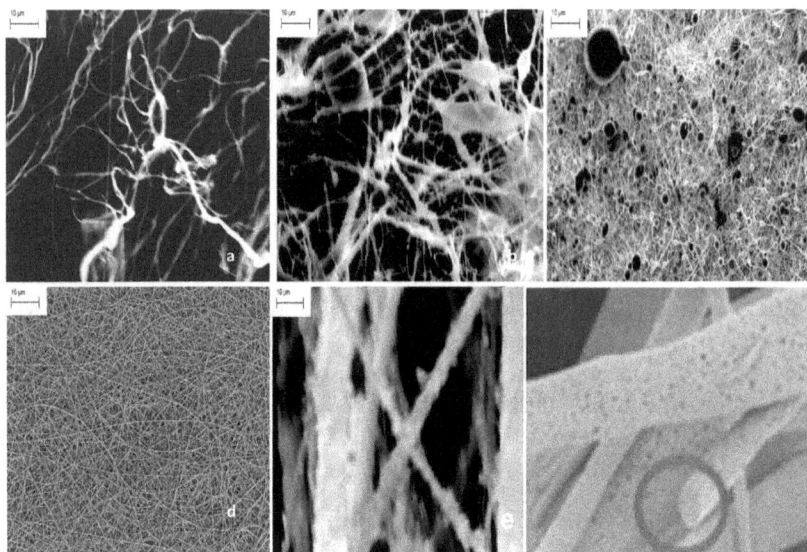

FIGURE 6.7 FESEM images of chitosan web with different solution concentrations: (a) 2%, (b) 5%, (c) 8%, (d) 11%, and (e) 14%. (f) Cross-sectional image of coated web.

the coated web could be produced in bulk at a substantially higher rate (about 0.25 $g/m^2/min$) than before, improving its economic feasibility. Overall, this study shows that needleless continuous electrospinning coating techniques can successfully electrospin and coat pure chitosan. The ideal electrospinning conditions were found, and the produced nanofibers underwent a detailed analysis. The chitosan nanofibers had different morphological features that ranged from dense to loose, and they were free of flaws. The formation of the coated webs was greatly aided by viscosity, and chitosan's existence was confirmed by chemical and crystalline characterisation. The resulting coated webbing has potential uses in antibacterial, filtration, wound healing, and dressing since it was air-permeable and porous. This study not only explains how very viscous materials can be transformed into practical coated nano-textile structures, but it also offers recommendations for manufacturing nanofibers from various bio-polymers based on polysaccharides. However, as the solution can get too thick to be electrospun, a balance between the viscosity and the polymer in the solution was required. The study also created an electrospun chitosan-coated nanofiber production technique that is very prolific and has the potential to be commercialized. The purpose of further investigation will be to monitor the development, effectiveness of filtration against germs and viruses, and wound healing procedure utilizing the chitosan-coated web structure [17, 18]. In comparison to traditional coated fabrics, the resulting web characteristics were significantly better at bacterial filtration and anti-bacterial effectiveness, as well as being lighter (Figure 6.8).

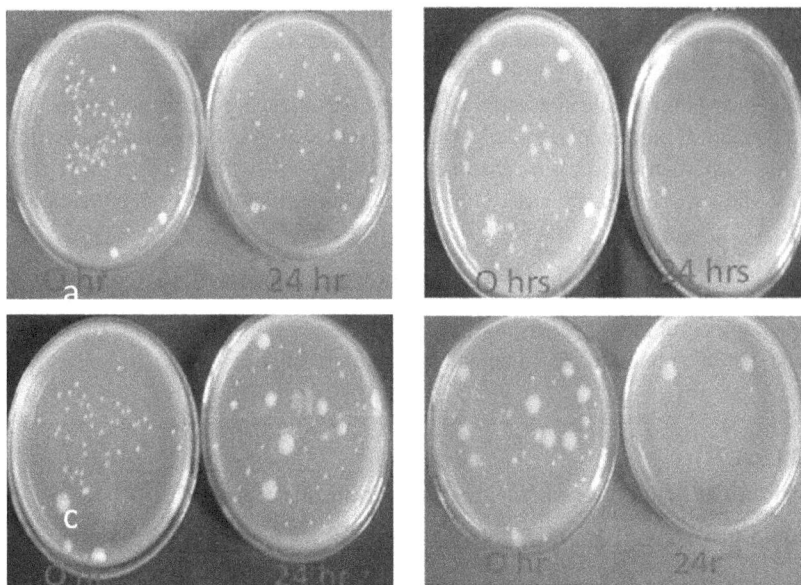

FIGURE 6.8   Agar plate images showing the effect of chitosan electrospun nanofiber webs on the bacterial incubation after 24 hrs (a and c). Uncoated *E. coli* (b and d) coated *E. coli*.

6.8.2   NANOFIBROUS POLYVINYL ALCOHOL WEB

Numerous factors affect the shape of electrospun nanofibers. This study examines how different process variables, including voltage, spinning electrode speed, electrode spacing, and solution concentration, affect the shape and tensile characteristics of electrospun nanofibrous PVA webs produced by a Nanospider machine [3]. According to FESEM pictures, which show the web density and nanofiber diameter, morphological changes are analyzed, and the mean tensile stress and Young's modulus are used to measure the tensile characteristics. Understanding how to control the process for the best growth of nanoweb morphology is made feasible by examining the unique influence of each parameter in isolation. The most efficient set of process variables for creating stronger nanofibrous webs is determined by correlating morphological changes with tensile properties in the study. By increasing the applied voltage, one can produce finer fibers by increasing the density of nanofibrous webs and the size of their pores. The spinning electrode speed affects the mass throughput rate, with greater speeds producing thicker, lower-density webs. A greater electrode distance results in coarser fibers, but a lower distance may cause web flaws. For applied voltage and electrode distance, the relationship between web density and fiber diameter is the polar opposite. The most crucial element is solution concentration because solvent viscosity impacts the merging behavior of the jets. Significant flaws in the nanowebs can result from insufficient solvent concentration. The density of the webs has a significant impact on their tensile characteristics, with higher density resulting in higher mass stress and Young's modulus.

The shape and tensile characteristics of the webs are significantly influenced by the production process parameters. Table 6.2 and Figure 6.9 show that by optimizing these characteristics, it is feasible to produce webs with high density, smaller holes, and superior tensile strength, making them perfect for filter application.

6.8.3 Electrospun Fabric for Ultraviolet Radiation Protection

By using a combined electrospinning and electrospraying technique, a protective fabric can be created to shield against ultraviolet radiation [11].

TABLE 6.2
Mechanical Properties of Various Experimental Trials

Process Parameters	Process Variables	Mean Tensile Stress (MPa)	Mean Young's Modulus (MPa)
Voltage (kV)	40	2.1	39.95
	80	6.9	107.16
Rotation of electrode (rpm)	6	7.8	126.43
	14	2.2	44.18
Spinning distance (mm)	110	2.9	56.87
	150	8.8	148.99
Solution concentration (w/v%)	3	8.1	150.87
	6	7.2	130.19

Effect of voltage 40 KV and 80 KV (AD 850 & 210 nm)

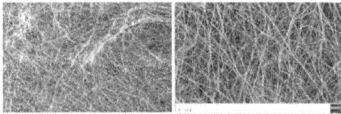

Effect of spinning electrode 6 rpm & 14 rpm (AD 180 & 520)

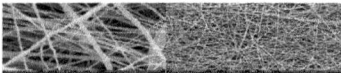

Effect of distance 110 & 150 mm (AD 820 & 320nm)

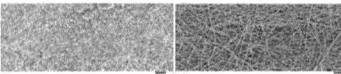

Effect of solution concentration % 3 & 6 (AD 310 & 425 nm)

Effect of	Variables	Process variable combinations
Voltage (KV)	40-80 (Intervals of 10)	6 w/v %, 150 mm, 14 rpm
	AD 850, 670, 540, 430, 210 nm	
Rotation electrode (rpm)	6-14 (intervals of 2)	6 w/v %, 150 mm, 80 kV
	AD 180, 240, 385, 490, 520 nm	
Spinning distance (mm)	110-150 (intervals of 10)	6 w/v %, 14 mm, 80 kV
	AD 820, 712, 610, 415, 320 nm	
Solution concentration (w/v %)	3-15 (intervals of 3)	150 mm, 14 rpm, 80 kV
	AD 310, 425, 690, 855, 1110 nm	

FIGURE 6.9   Effect of process parameters.

The method for producing UV radiation-resistant fabric described in the article combines electrospinning and electrospraying processes. The researchers developed a nanofibrous web of PVDF-coated polypropylene nonwoven fabric using the most recent nanospider technology. They then functionalized the web with titanium dioxide (TiO_2) to improve its UV protection properties. The resulting web was evaluated for its surface morphology, mechanical, chemical, crystalline, and thermal properties. The researchers achieved an optimal nanofiber spinning condition to produce a defect-free functional nanofiber coating on the web, which could be used to create lightweight and protective clothing, tents, covers, and shelters for defense, paramilitary, and civilian use. The functionalized web had a UV protection factor of 65, compared to 24 for the non-functionalized web. Results from electron microscopy and field emission scanning electron microscopy demonstrated that the photo-degradative behavior of the PVDF webs was non-accelerated, and TiO_2 functionalization improved the web's photo-stability (Figures 6.10 and 6.11).

The purpose of this research is to gain a scientific understanding of the photo-degradation properties of PVDF fiber-coated webs and their potential applications in protective textiles for defense, including segments, combat suits, and snowbound camouflaging nets. The goal is to conduct laboratory-scale research that can be further developed into large-scale production.

Using the most recent nanospider machine, the research endeavor was successful in fabricating, optimizing, and evaluating the functional properties of PVDF-coated and functionalized webs. The absence of flaws like holes, stains, knots, or beads in the finished fiber web demonstrates how well the production process worked. As shown in Figure 6.12, the PVDF web demonstrated non-accelerated photo-degradative behavior against accelerated weathering circumstances, which may be attributed to the material's inherent inertness.

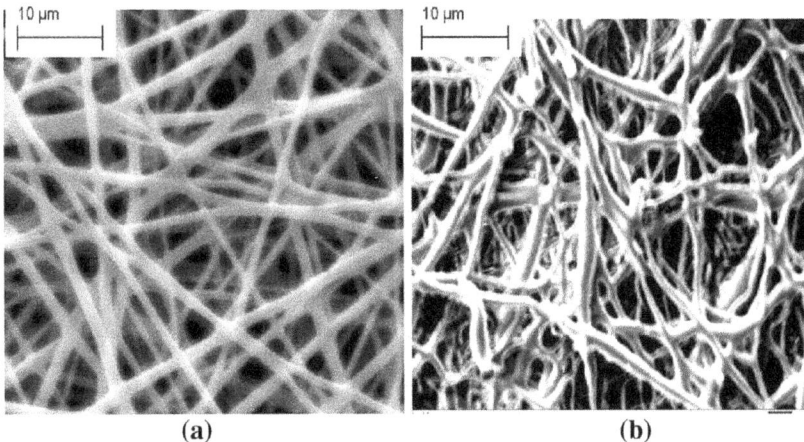

(a) **(b)**

FIGURE 6.10 FESEM images of PVDF nanofibers coated on PP nonwoven fabric: (a) nonfunctionalized web and (b) functionalized web with TiO_2 nanoparticles.

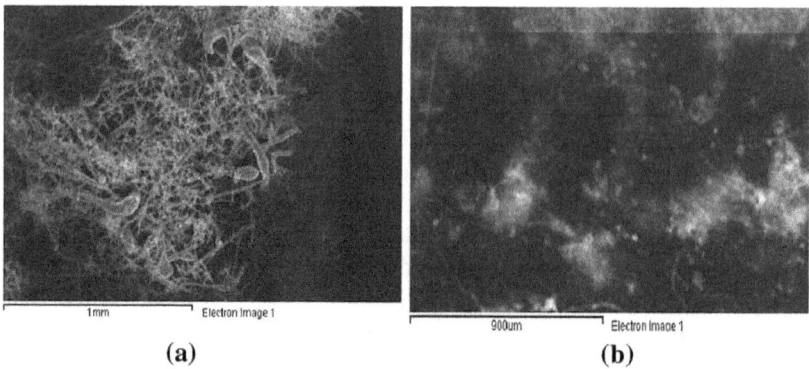

FIGURE 6.11 Electron images of PVDF nanofibers coated on PP nonwoven fabric: (a) nonfunctionalized web and (b) functionalized web with TiO_2 nanoparticles.

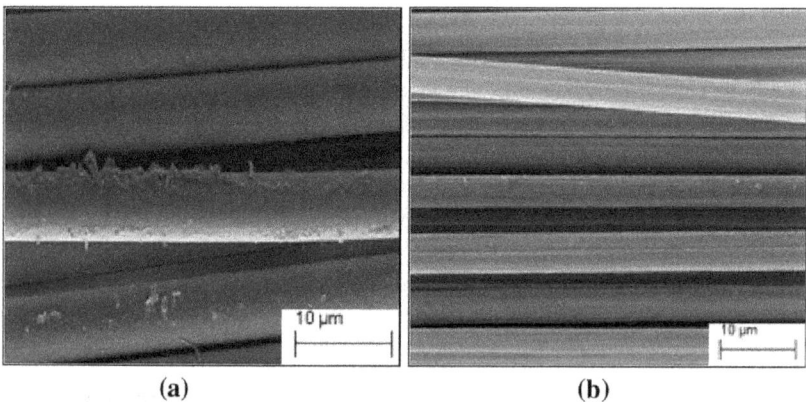

FIGURE 6.12 FESEM microphotographs of the exposed (100 h) PVDF nanofibrous web: (a) nonfunctionalized and (b) functionalized.

Figure 6.13 reveals the various structured nanowebs for technical textile applications. The functionalized web, on the other hand (Figure 6.13), demonstrated improved protective behavior against photo-tendering. This study shows how electrospun PVDF webs, which might be used as a useful layer in protective fabrics, can be optimized and made feasible. According to the research, the UPF of the functionalized nanowebs increased from 24 to 65, making them more effective at shielding against dangerous UV rays [11]. Understanding the behavior of materials in other applications, such as industrial fabrics used for outdoor applications (e.g., sportswear; protective fabric against skin cancer; and UV-protective fabrics such as tents, covers, shelters, aerostats, combat suits, snow-bound camouflaging fabrics, and nets), is made easier by the knowledge gained from this study. Additionally, the research could be expanded to create lightweight

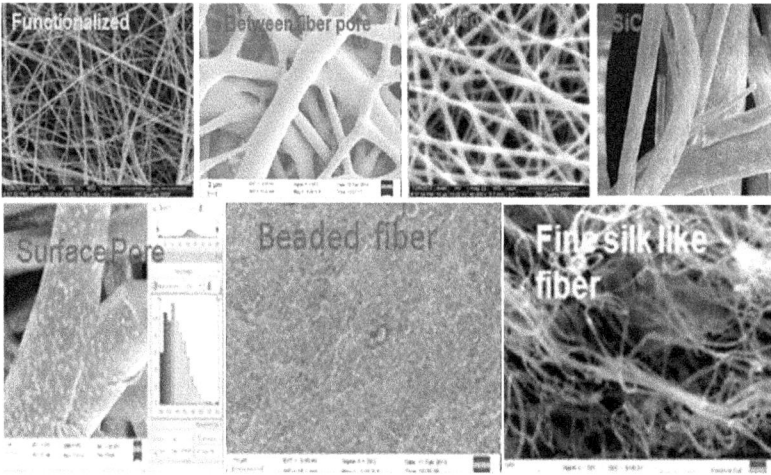

FIGURE 6.13   FESEM images of the different morphological structures of coated webs.

multifunctional webs with qualities including semi-permeable membranes (offering comfort and porosity), water repellency, UV protection, flame retardancy, and bacterial filtering.

6.9   CONCLUSION

This chapter provides an overview of the current status of nozzle-less technology (via spinning electrodes) for producing nanofiber webs. It explains the underlying principles, operations, equipment, capabilities, and key features of the technology. The chapter also discusses the general applications of nanofiber webs and highlights the research efforts and practical utilization of the technology in various technical textile sectors. The focus of the chapter is on the practical aspects of using nanofiber webs for bulk production and commercialization of web products. Nanofibre web coatings have tremendous potential for creating lightweight textiles with multiple functionalities, including water repellency, conductivity, high strength, protection against hazardous chemicals and warfare agents, improved bacterial and biological filtration efficacy, UV protection, and flame retardation. These specialized fabrics can also be used for semi-permeable membranes, wound healing, bacterial filtering, and medication delivery.

REFERENCES

1. Lim CT. 2017. Nanofiber technology: Current status and emerging developments. Progress in Polymer Science 70:1–17. https://doi.org/10.1016/j.progpolymsci.2017.03.002

2. Esthappan SK, Sinha MK, Katiyar P, Srivastav A, Joseph R. 2013. Polypropylene/zinc oxide nanocomposite fibers: Morphology and thermal analysis. Journal of Polymer Materials 30(1):79–89.

3. Sinha MK, Das BR, Srivastava A, Saxena AK. 2013. Needleless electrospinning and coating of poly vinyl alcohol with cross-linking agent via in-situ technique. International Journal of Textile and Fashion Technology 3(5):29–38.

4. Sinha MK, Das BR, Mishra R, Ranjan A, Srivastava A, Saxena AK. 2014. Study of electrospun polycarbosilane (PCS) nanofibrous web by needle-less technique. Fashion and Textiles 1(1):1–14. https://doi.org/10.1186/s40691-014-0002-9

5. Shen L, Yu X, Cheng C, Song C, Wang X, Zhu M, Hsiao BS. 2016. High filtration performance thin film nanofibrous composite membrane prepared by electrospraying technique and hot-pressing treatment. Journal of Membrane Science 499:470–479. https://doi.org/10.1016/j.memsci.2015.11.004

6. Bocanegra RS. 2021. Development of new strategies for the design of in situanalysis devices: Nano and biomaterials (Thesis). http://roderic.uv.es

7. Sinha MK, Das BR, Srivastava A Saxena AK. 2014. Study of electrospun chitosan nanofibrous coated webs. Journal of Nano Research 27:129–141. https://doi.org/10.4028/www.scientific.net/JNanoR.27.129

8. Barhate RS, Ramakrishna S. 2007. Nanofibrous filtering media: Filtration problems and solutions from tiny materials. Journal of Membrane Science 296(1–2):1–8. https://doi.org/10.1016/j.memsci.2007.03.038

9. Kim C, Yang KS. 2003. Electrochemical properties of carbon nanofiber web as an electrode for supercapacitor prepared by electrospinning. Applied Physics Letters 83(6):1216–1218. https://doi.org/10.1063/1.1599963

10. Sinha MK, Das BR, Kumar K, Kishore B, Prasad NE. 2017. Development of ultraviolet (UV) radiation protective fabric using combined electrospinning and electrospraying technique. Journal of The Institution of Engineers (India): Series E 98(1):17–24. https://doi.org/10.1007/s40034-017-0094-z

11. Li W, Li X, Wang Q, Pan Y, Wang T, Wang H, Song R, Deng H. 2014. Antibacterial activity of nanofibrous mats coated with lysozyme-layered silicate composites via electrospraying. Carbohydrate Polymers 99:218–225. https://doi.org/10.1016/j.carbpol.2013.07.055

12. Kim BH, Yang KS, Ferraris JP. 2012. Highly conductive, mesoporous carbon nanofiber web as electrode material for high-performance supercapacitors. Electrochimica Acta 75:325–331. https://doi.org/10.1016/j.electacta.2012.05.004

13. Chavan S, Kanu NJ, Shendokar S, Narkhede B, Sinha MK, Gupta E, Singh GK, Vates UK. 2023. An insight into nylon 6, 6 nanofibers interleaved E-glass fiber reinforced epoxy composites. Journal of the Institution of Engineers (India): Series C 104(1):15–44. https://doi.org/10.1007/s40032-022-00882-0

14. Sinha MK, Das BR. 2018. Chitosan nanofibrous materials for chemical and biological protection. Journal of Textiles and Fibrous Materials 1:1–13. https://doi.org/10.1177/2515221118788370

15. Dasaradhan B, Das BR, Sinha MK, Kumar K, Kishore B, Prasad NE. 2018. A brief review of technology and materials for aerostat application. Asian Journal of Textiles 8(1):1–12. https://doi.org/10.3923/ajt.2018.1.12

16. Celebioglu A, Uyar T. 2020. Fast-dissolving antioxidant curcumin/cyclodextrin inclusion complex electrospun nanofibrous webs. Food Chemistry 317:126397. https://doi.org/10.1016/j.foodchem.2020.126397

17. Sinha MK, Das BR, Prasad N, Kishore B, Kumar K. 2018. Exploration of nanofibrous coated webs for chemical and biological protection. Zaštita Materijala 59(2):189–198. https://doi.org/10.5937/ZasMat1802189K

18. Dai F, Huang J, Liao W, Li D, Wu Y, Huang J, Long Y, Yuan M, Xiang W, Tao, F, Cheng Y. 2019. Chitosan-TiO2 microparticles LBL immobilized nanofibrous mats via electrospraying for antibacterial applications. International Journal of Biological Macromolecules 135:233–239. https://doi.org/10.1016/j.ijbiomac.2019. 05.145

7 Future Perspectives on High-Performance Technical Textiles

This book provides a detailed discussion on fiber to fabric technology, including a classification and global overview of the industry. It also highlights the research gap in the use of functional and high-performance fibers for high-value technical textile applications. Practical product-oriented research is presented, with relevant examples that meet stringent technical textile quality standards. In technical textile products, functional fibers are the best option. The concept of moisture management in clothing is essential for providing maximum comfort even in high-humidity conditions. To achieve this, further research is needed to establish weaving and processing parameters for technical textile products. Such efforts will pave the way for replacing multi-layered technical textiles with a single-layered multifunctional structured fabric, eliminating the need for tedious fabric finishing and coating operations. Technical textiles are produced globally using high-performance or functional fibers, resulting in improved functional properties and lighter-weight materials. These specialty fibers and composites include composites made from fibers with specific applications, the development of conductive yarns, and bi-component fibers that maintain permanent crimp and have been studied for their thermal insulation properties. Additionally, micro-denier fibers have been developed using different polymers and cross-sectional shapes. Technical textiles have also been developed as substitutes for plastics and metals to promote sustainability, recycling, and trackability in the textile value chain. Other technical textile innovations include breathable PU-coated lightweight fabrics, UV-reflective coated fabrics, NIR-reflective printed PU-coated fabrics, and pile acrylic fabrics with polyester base and pile fibers of 1–3 denier.

These technical textiles encompass high-performance fibers, yarns, woven, knitted, nonwoven, braided, and composite structures. An important area of research in technical textiles is the development of 3D fabric structures, such as interlock woven or multi-axial warp knit, using high-performance fibers. The focus is on understanding the impact of various manufacturing parameters, including weave geometries, filling density, stitch length, and yarn orientation, on the stiffness, strength, energy absorption, fracture behavior, and other performance parameters of composite panels made with standard resins. Empirical relationships between constructional variables and performance parameters are being identified and organized into a data bank. Additionally, research on functional fibers and inflatable textiles for high-end uses will pave the way for developing futuristic

DOI: 10.1201/9781003317074-7

technical textile segments. The practical and feasible applications of technical textiles, from fiber to manufacturing stages, are also highlighted in these chapters to demonstrate how focused research can be translated into product development at an industrial scale rather than remaining solely of academic interest.

Gel-spun polyethylene has numerous applications in various fields such as sailcloth, marine ropes and cables, impact shields, body armor, medical implants, sports equipment, boat hulls, ballistic protection, concrete reinforcements, and fishing nets, among others. Its exceptional property for sailcloth is its high creep resistance, which ensures minimal distortion for a lightweight, high-strength sail. In the case of marine ropes, gel-spun polyethylene boasts a unique combination of properties such as light weight, high strength, excellent abrasion resistance, low moisture absorption, and inert behavior in saline water, making it a highly competitive product compared to aramids. It has been reported that extended chain polyethylene is eight times better than aramid fiber in cyclic loading and wear tests, with its abrasion resistance also being eight times better than that of aramids. The material's light weight, high strength, modulus, and good impact strength make it a viable option for composites and other high-performance applications.

Protective clothing for fireman, para-military forces, traffic, sanitation, industry, and security personnel covers garments and accessories able to protect them from hazardous materials and processes in the workplace. Such textiles provide protection from hail, water, and fire and yet maintain the breathability and lightness of the textile materials.

The advantages of agricultural by-products in technical textiles and finishes are a boon to the environment, industry, and individuals that can not be ignored in the current sustainable scenario. The materials and products sourced from agro-residue are also farmer friendly, solving their problem of residue disposal.

Chapter 6 provides a comprehensive analysis of the potential of nanofibrous materials, created through needle-less technology, for producing lightweight functional technical textiles. These futuristic materials include Nylon 66, high-strength polyester, polyacrylonitrile (PAN), carbon nanotube (CNTs), PAN/CNT composite fibers, polybutadiene-styrene rubber (PBSR), activated carbon nanofibers (ACNFs), polycarbosilane, and composites materials for silicon carbide composite fibers, chitosan webs, functionalized nanowebs with inorganic nanoadditives and pigments for detoxification, and other applications.

These high-performance materials are spun using needle-less nanospider technology and coated onto nano-woven textile fabrics. The resulting coated fabric possesses a range of functional properties, such as flame retardancy, UV protection, high porosity, resistance to toxic chemicals, ferromagnetic properties, antimicrobial/fungal and antibacterial properties, higher bacterial filtration efficiency, ballistic performance, and water and oil repellency. Compared to traditional textiles, these materials offer numerous advantages, including a higher degree of functional properties due to their larger surface area per square meter, even at minimal coating densities ($0.2–1.8$ g/m^2). Research has demonstrated that these higher-performance materials can withstand rugged conditions without damaging the fiber structure. Furthermore, they have a longer usable life, are highly practical, and utilize scalable technology.

An inherently multifunctional FR polyester woven (plain) fabric is suitable for use in various technical textiles. The fabric is made using multifilament threads and has a mass density of approximately 140 gsm. The thread properties and fabric construction parameters are specified. After production, the finished FR fabric is assessed for various functional properties such as flame retardancy, oil repellency, water repellency, air permeability, and anti-static behavior.

Although there is some flame debris in the weft direction, the fabric meets the desired flame retardancy requirements. Additionally, the fabric has excellent oil repellency, with no instances of penetration observed during testing. The water repellency rating was measured at 90 for both normal and washed conditions. The air permeability of the finished fabric was assessed to be 59.8 and 54.7 cc/sec/cm^2 for normal and washed conditions, respectively. The surface resistivity of the fabric was found to be 3×10^{-9} Ohm, which is adequate for dissipating any static charge that may develop.

Industrial collaboration is crucial for scaling up productization efforts. Furthermore, the technical textile industry requires a multidisciplinary approach, involving not just textile technologists but also experts in other relevant fields.

Index

Note: Page numbers in *italics* indicate figures, and page numbers in **bold** indicate tables in the text.

For Product Safety Concerns and Information please contact our EU
representative GPSR@taylorandfrancis.com
Taylor & Francis Verlag GmbH, Kaufingerstraße 24, 80331 München, Germany